Local Government Area boundaries, Sydney, 1990

Taken for Granted:
The bushland of Sydney and its suburbs

Flowering *Xanthorrhoea*, Grass Trees, are striking in the diverse understorey of Mowbray Park's hillside open-forest, one of several plant communities in this park on the banks of the Lane Cove River at Lane Cove West. (J. Plaza, RBG, 1990)

Taken for Granted: The bushland of Sydney and its suburbs

Doug Benson and Jocelyn Howell

Aboriginal life on the shores of Port Jackson as recorded by the French artist Lesueur in 1802. 'In these canoes they will stand up to strike fish, at which they seem expert,' Lieutenant William Bradley had written in 1788. (Charles Alexandre Lesueur, 1778–1846, *Grottes, chasse et peche des sauvages du Port-Jackson*, Rex Nan Kivell Collection (4539), National Library of Australia)

Kangaroo Press

in association with the

Royal Botanic Gardens Sydney

Acknowledgments

Many people have contributed to this work, both in the long-term gathering of information, and in the more immediate process of preparing a book. We would like to take this opportunity to thank people who have, over past years, helped collect information used as the basis of the book, particularly former Ecology Section staff Helen Bryant, David Keith, Richard McRae, Martin Cooper, Heidi Fallding, Jeff Thomas, Jane Burkitt and Peter Clarke. Many others, too numerous to name, have contributed comments and information on specific areas amongst them we would like to thank Paul Adam, Don Adamson, Heather Adamson, Robin Buchanan, Colin Gibson, Roger Lembit, Eddie McBarron, Lyn McDougall, Astrid Mednis, Robert Miller, Louisa Murray, Peter Myerscough, Chris Pratten, Tony Price, Graham Quint, Malcolm Reed, Sue Rose, Peter Smith, Gay Spies and David Thomas. We particularly appreciate the effort taken by Chris Pratten, Louisa Murray, Barbara Briggs and Lawrie Johnson in reviewing the entire text.

We are indebted to Peter Mitchell for his major contribution to Chapter 1, and to Jim Kohen, who generously provided information for Chapter 2. Our special thanks go to Lyn McDougall for dedication beyond the call of duty in producing the maps, and to Jaime Plaza for cheerful and generous contribution of his photographic skills. We are very grateful to Elizabeth Brown for giving her time to draw the geomorphological diagram.

We would like to acknowledge the help of all staff of the National Herbarium who have provided advice on plant groups or areas of vegetation known to them, particularly Bob Coveny, Gwen Harden, Teresa James, Barbara Wiecek, Seanna McCune, Bob Makinson and Barry Conn.

We also thank all Royal Botanic Gardens staff who contributed to typing the manuscript, and particularly Cindy Douglas, Pauline Shires, Angela Benn and Sue McCahon for their cheerful and longsuffering contributions.

For providing key information on request we thank Ken Robinson, Alf Salkin, Brian Wray and Harley Wright. Sue Johnston, Helen McDonald, Megan Martin, Fred Midgley and David Russell provided valuable assistance in obtaining historical photographs of their local areas, and our special thanks go to Nelson Blissett and Phyllis Schaffer for kindly lending their own photographs. Our thanks also to Liz Bird, Leanne Collins, Leigh McCawley, Ron Peck and Catherine Snowden for help with illustrations.

We are very appreciative of the support of Professor Carrick Chambers and Dr Barbara Briggs during the creation of this book.

Jacket design by Darian Causby using the painting:
Sydney Heads 1865
Eugene von Guerard
Oil on canvas 56 × 94 cm
Bequest of Major H.W. Hall, 1974.
Collection Art Gallery of New South Wales

Benson, D. H.
Taken for granted : the bushland of Sydney and its suburbs.

Bibliography.
Includes index.
ISBN 0 86417 331 8.

1. Botany - New South Wales - Sydney Region. I. Howell, Jocelyn, 1944- . II. Royal Botanic Gardens (Sydney, N.S.W.). III. Title.

581.99441

First published in 1990 by Kangaroo Press Pty Ltd
3 Whitehall Road (P.O. Box 75) Kenthurst NSW 2156
Typeset by G.T. Setters Pty Limited
Printed in Singapore by Fong & Sons Printers Pte Ltd

Contents

Introduction

At the beginning of 1788 the bushland of Sydney — an uninterrupted mosaic of forests, woodlands, heaths, scrub, sedgelands and swamps — stretched from the coast, west to the Nepean-Hawkesbury River. The various patterns were a response to the underlying geology and soils, modified by the effects of climate, particularly rainfall and temperature. Forests occupied the most fertile and well-watered lands — those close to the coast or associated with the rich floodplains of the Nepean-Hawkesbury River. Grassy woodlands spread across the clay soils of the drier Cumberland Plain, shrubby woodlands covered the poorer sandy soils of the Hawkesbury Sandstone ridges. Heaths and scrub occupied shallow, sandy soils or very exposed coastal sites. Swamps filled poorly drained depressions, and mangroves and saltmarsh fringed sheltered coastal estuaries.

This landscape had been occupied and modified by Aboriginal peoples for thousands of years. By 1888, a century after European settlement's shaky start on the infertile soils of Farm Cove, agriculture had changed irretrievably the native vegetation on a previously 'unproductive' continent and been responsible for the destruction of much of its flora and fauna.

On Sydney's Cumberland Plain the impact on native plant and animal species has been devastating. Clearing and cultivation removed bushland completely. Changes in burning frequencies and intensities as Aborigines were displaced, and new grazing pressures as domestic stock replaced the native marsupials, altered bushland irrevocably. There is such scant knowledge of the original vegetation diversity that it is hard to describe the magnitude of these changes. These impacts were to be repeated elsewhere in New South Wales during the following century and right up to the present, but the effects of agriculture on Sydney's bushland, except for hobby farm development which is essentially suburban, had ceased by the end of the nineteenth century.

The story of Sydney's bushland during its second century of European occupation is essentially a story of expanding suburbs and consequent bushland retreat. Perhaps prophetically, the first trees cut in Sydney Cove in 1788 were felled to make way for huts. Sydney started to acquire its suburban character in the 1880s as the wealth acquired in the gold rushes and from agriculture, led to an expanding population and housing boom. The development of transport networks with trains, trams, ferries and buses, and later private cars, promoted the spread of the suburbs. After more than a century, this relentless process continues.

Sydney's bush is now 'suburban bush', largely confined to creeks or steep slopes, in spaces unsuitable or uneconomic for housing or at some stage unavailable because of defence or other needs. Certainly the major national parks to the north, west and south have very large natural bushland areas, but, like the small reserves, their major management problems are of suburban origin; weed invasion from upstream residential areas, rubbish dumping, changes in fire regimes along suburban interfaces, theft of bush rock and Grass Trees *(Xanthorrhoea)* for suburban gardens, and the effects of large concentrations of visitors.

The future for Sydney's bushland therefore lies in understanding the relationship between the suburb and the bush; in understanding how particular suburban practices will have inevitable bushland consequences, how nutrients in stormwater run-off allow weeds to enter the bush and how changes in burning patterns change the relative abundance of different species.

Sydney's bushland past was different from Sydney's bushland present, and will be different from Sydney's bushland future. Here we look at the past and present — and the implications for the future.

1 Sydney's Landscape

'a jumble of rocks and thick woods'

Geology

The landscape we know today began to take shape nearly 300 million years ago, during the geological period known as the Permian. Since then, tectonic forces associated with the movement of continents, rising and falling sea-levels, changing climates, violent but short-lived volcanic eruptions, and the continuing wearing processes of erosion, have all left their mark.

During the Permian Period, Sydney lay at the mouth of a broad swampy river basin, covered with lush plant life; ferns, gingkos, primitive conifers, seed ferns, horsetails and trees of *Glossopteris*. These plants included the ancestors of today's vegetation, and their fossils provide evidence of the former connection of the southern continents in a supercontinent called Gondwana.

About 230 million years ago, at the beginning of the Triassic Period, rivers eroding inland mountains began delivering enormous quantities of sand, silt and clay to the coast. These sediments accumulated on the floodplains and in deltas on a scale comparable to the present Nile or Ganges Rivers. Successive layers of sediment filled the Sydney Basin, burying the Permian swamps and turning their organic remains into coal layers (the Illawarra and Newcastle coal measures). Hundreds of metres of sandy sediments were cemented into sandstone, and the finer silts and clays to mudstone and shale. Ultimately the lower layers became today's Narrabeen Group sandstones and shales, the sandy middle beds, the Hawkesbury Sandstone, and the uppermost (youngest) strata of compressed silt and clay, the Wianamatta Shale.

By the end of the Triassic, about 190 million years ago, sediment accumulations in the Sydney area had almost stopped. Australia was still part of the supercontinent Gondwana, connected through Antarctica to India, Africa and South America, but pressures building up within the earth's crust had begun to split the supercontinent apart. In the Sydney Basin the initial movements were marked by explosive eruptions of small volcanic breccia pipes (diatremes) through the sedimentary rocks, and by the injection of dolerite into the Wianamatta Shale at Prospect. Then over the next 140 million years, during the Jurassic and Cretaceous and into the early Tertiary Periods, the continents slowly separated, and moved toward their present positions. Africa and South America separated early from Antarctica. Australia, still joined to Antarctica, moved slowly northwards, undergoing final separation about 40 million years ago.

As continental plates moved further apart to form the Tasman Sea, there was highland uplift and volcanic activity in south-eastern Australia. Deep-seated crustal pressures under Sydney caused the Blue Mountains to rise and the Cumberland Plain to be slowly lowered along a line of flexure (the Lapstone monocline and the Kurrajong fault). This line is followed by the modern Nepean-Hawkesbury River. Earth movements were very slow — so slow that erosion of the rivers kept pace with tectonic movements — and over 50 million years or more of the Tertiary Period, the rivers cut deep gorges through the erosion-resistant sandstone that remained as higher plateaus near the coast. The ancestral Nepean-Hawkesbury River, flowing across the Cumberland Plain, laid down alluvial sediments — gravels, clays, silts, and sand. These formed a series of terraces in the Windsor–Castlereagh district and, near Liverpool, similar smaller terraces formed along the Georges River. These sediments, variously

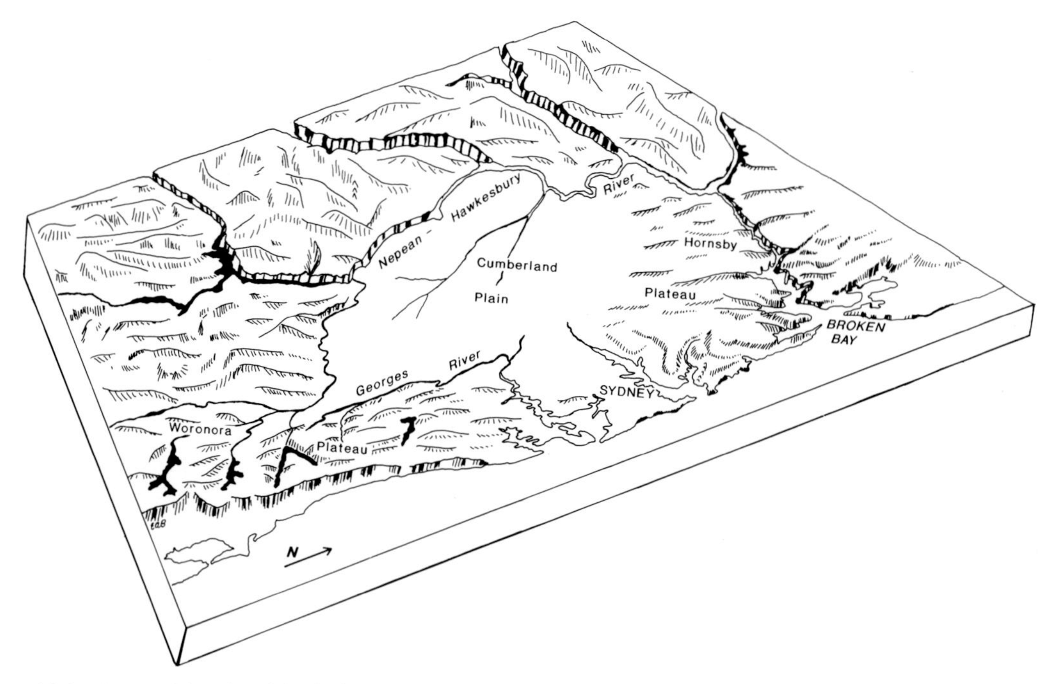

Major geomorphic units of the Sydney area.

weathered, are known as Tertiary alluvium; some have characteristic ironstone concentrations.

The most recent stages in the development of the Sydney landscape occurred during the last 1.8 million years (in the Quaternary Period). Alluvium continued to accumulate on the flood plains and terraces of all the main streams; particularly the Nepean-Hawkesbury near Camden, Penrith and Windsor, along South Creek, Rickabys Creek and Eastern Creek, and along the Georges and Parramatta Rivers. Dune sands were blown inland from the coast where today's southern and eastern suburbs lie. Colder and warmer periods alternated. About 20,000 years ago, during the coldest part of the last of the Pleistocene ice ages, the sea fell to its lowest level, 120–140 m below the present. As it rose again to reach its present level about 6,000 years ago, it drowned the coastal river valleys to form Port Hacking, Sydney Harbour, Pittwater and Broken Bay, and swept up the offshore sands on to the modern beaches, sometimes damming smaller streams to form lagoons (such as at Narrabeen, Dee Why and Wattamolla).

The Sydney Basin sediments, deposited over millions of years, now form a broad, slightly lopsided saucer, cut off on the eastern side by the Pacific Ocean. At its edges, the coarse-textured Hawkesbury Sandstone that characterises much of Sydney's landscape rises over 200 m, forming the Hornsby Plateau in the north, the Woronora Plateau in the south, and the Blue Mountains Plateau in the west. Underlying the Hawkesbury Sandstone are the upper strata of the Narrabeen Group, with their interbedded layers of shales, sandstones and claystones, outcropping along the coast north of North Head, and to the south in Royal National Park. Below the Narrabeen Group, but too deep to outcrop at the surface, lie the Illawarra and Newcastle coal measures. Coal was mined from pits over 800 m deep in the harbourside suburb of Balmain in the early years of this century.

Overlying the Hawkesbury Sandstone, but now mostly topographically lower as a result of the subsidence of the Cumberland Plain, are the shales and occasional sandstones of the Wianamatta Group. These shales are found over a large part of the inner western and southern suburbs, and extend across Sydney's west and south-west as far as Windsor, Penrith, Picton and Campbelltown. Thin layers of Wianamatta Shale also cap the sandstone ridges on the North Shore.

Sydney's largest body of igneous rock is at Prospect Hill, formed about 170 million years ago in the Jurassic Period, when molten rock intruded through the sandstone and into the shale, cooling before reaching the surface, to form dolerite. As well,

30 or so small diatremes, collapsed volcanic 'vent-holes', containing volcanic breccia and/or basalt (often mined for 'blue metal'), are scattered throughout the area. The breccias erode faster than sandstone but at a similar rate to shale. Thus in sandstone areas diatremes occur as eroded depressions or valleys, often covered with alluvium or colluvium. The Hornsby diatreme, in Old Mans Valley, is the largest. In shale country the diatremes formed hills, but most, as at Dundas, Minchinbury and Erskine Park, have now been quarried.

Soils

Despite a limited range of rock types, Sydney's soils vary considerably, because factors controlling soil type also include local climate, position on slope, and interactions with plants and animals.

The Hawkesbury Sandstone landscapes share a topography of steep hills, long narrow ridges (sometimes capped by remnants of Wianamatta Shale) and deep rocky valleys. They also have shallow, sandy, infertile soils because the weathering products of the quartz sandstone are little more than quartz sand. Though this material provides a considerable challenge to the home gardener, the native flora of these areas has a surprising variety.

In the sandstone areas four different types of soil may be found. On some of the broader ridges and plateaus, weathering of the rock may proceed undisturbed for a long time, and a yellow earthy profile 1 to 2 m deep can develop (for example at Terrey Hills). These yellow earths are of limited extent, and although they contain less than optimal nutrients for plant growth, they are well drained and have a structure that allows good root penetration.

At the edge of the plateaus and on the rugged sandstone slopes, between rock outcrops, the soil profiles consist of two layers of material. The topsoil, invariably sandy and often containing many rock fragments, is a blanket of sediment that is kept more or less in constant motion downslope by erosion. The subsoil varies with the nature of the sandstone. A simple quartz sandstone weathers to single-grained sand, and the two layers of soil materials in this case produce uniform soil profiles of siliceous sands, or yellow or grey earthy sands. Where the sandstone contains some clay, or perhaps the topsoil is underlain by a shale bed, a clayey subsoil will develop and the profile will show a marked contrast in texture, with sandy topsoil over clayey subsoil. These profiles have been called red and yellow podzolics. Towards the foot of the slope and sometimes on benches on the slope, the surface blanket of eroding sands can accumulate to depths of 2 or 3 m, and on some of these sites a particular group of plant species can be found. These interact with the sand to produce a podzol, a rather unusual soil profile with a thick layer of bleached white sand just below the surface and cemented iron and/or organic layers (pans) in the subsoil.

Landscapes on the Wianamatta Shale and alluvium of the Cumberland Plain have a much gentler topography. Here the more fertile loamy or clay soils retain more moisture and have slightly higher nutrient levels, and many areas were used for market gardens and orchards. On upper slopes the soils are mostly red or brown podzolics, grading into yellow podzolics further down the hillside. These profiles are formed as two layers of material, in much the same way as the profiles on the sandstone but the contrast in texture between the loamy topsoil and the clay subsoil is less marked. On lower slopes concentrations of iron nodules may be found in the subsoils. Wianamatta Shale soils contain variable amounts of salt, depending on the extent of leaching that occurs under local rainfall and landform conditions. Salt content can be significant enough to affect plant distributions, particularly in low-lying areas. For example, *Casuarina glauca*, the Swamp Oak, normally associated with estuarine sites, grows in some low-lying drainage lines where salt leached from the surrounding shale soil has made conditions saline.

Soils derived from the volcanic rocks are more fertile than surrounding shale and sandstone soils. On the Prospect intrusion, well-structured, uniform, brown clay soils are found which have been classified as prairie soils and black earths. On some of the larger diatremes and dykes deep red clays called kraznozems occur. Even when the weathered rock itself is buried by debris from the sedimentary rocks on the adjacent slopes, local soil fertility may be higher than in surrounding soils.

The soils developed on the Tertiary and Quaternary alluvium vary with the nature of the sediment. Podzolics are common on the older, Tertiary clayey materials, loamy soils are found on the younger sediments, and podzols are found on some of the cleaner sand deposits (such as at Agnes Banks) and on the coastal dune sands. Immediately behind some beaches calcareous sands occur (Palm Beach, Mona Vale and the Kurnell Peninsula), but very simple profiles of siliceous sands are more common. Organic sands and muds are usually found

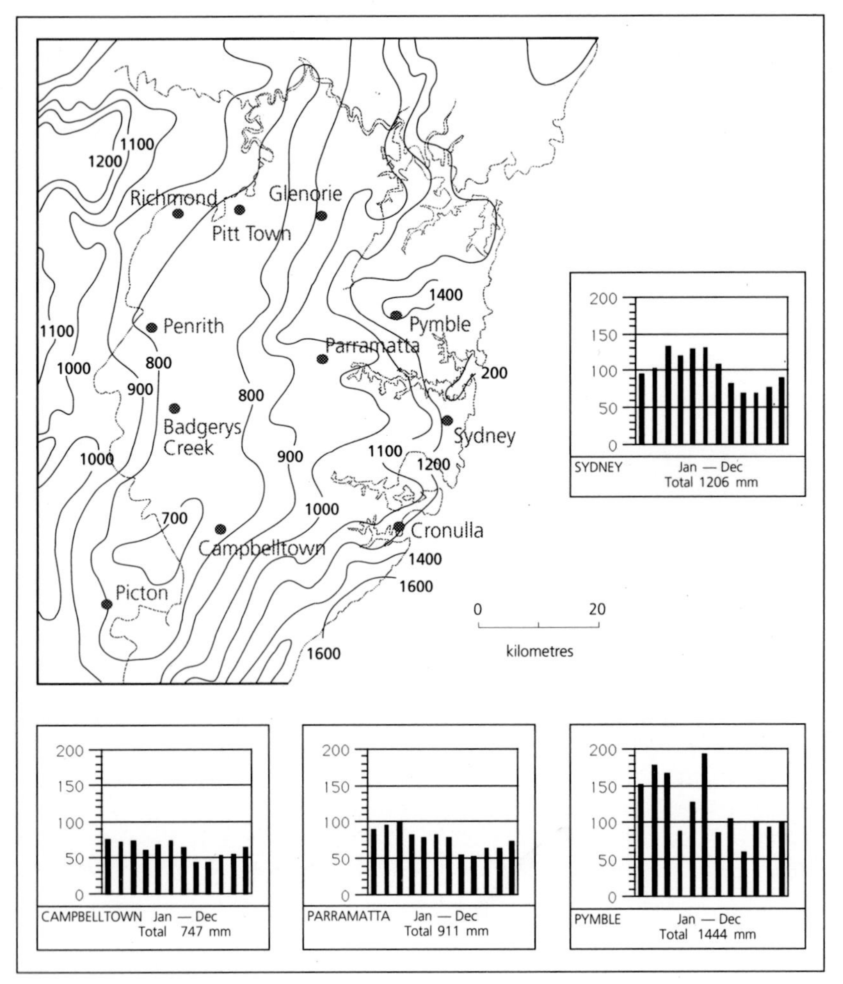

Sydney's average annual rainfall: isohyets in millimetres.

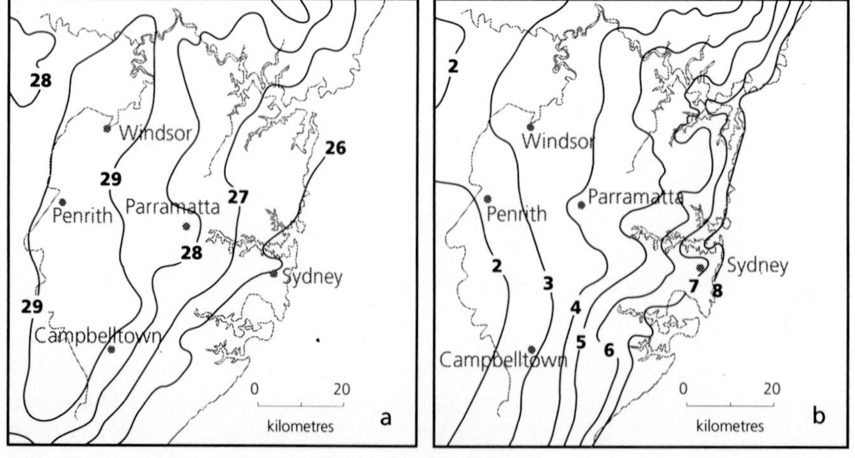

Sydney's temperature extremes: isotherms in degrees Centigrade: (a) average January maxima; (b) average July minima.

on the margins of the estuaries and coastal lagoons.

Phosphorus is a limiting nutrient in all of the soils of the Sydney area. The highest concentrations are found in the surface litter and organic layers except in the soils from volcanic rock where it is more uniformly distributed. Slightly more phosphorus is found in the soils derived from shales than in those derived from sandstone but the biggest differences in quality between these relate to soil structure and the ability of the heavier textured soils to retain more soil moisture. Almost all the soils in Sydney are acidic both at the surface and at depth, although again the volcanic soils and a few soils derived from carbonate cemented sandstones are an exception.

Within the older suburbs and the central business district many of the modern soils have been extremely disturbed and may even be artificial. Original topsoil has often been stripped and replaced by building debris and landfill materials, including garbage. Many sites are covered by compacted clay, layers of sand and grass turf. Samples of such disturbed soils from the site of the first Government House in Bridge Street are more alkaline, with much higher levels of organic carbon, calcium and phosphorus than in the undisturbed state.

Climate

Sydney is part of the subtropical east coast, and experiences a warm wet summer–autumn and cool drier winter–spring. Local conditions vary according to topographical features and distance from the sea. From the coast to the Cumberland Plain, rainfall follows a decreasing gradient, temperature extremes become more pronounced, and there is an increasing incidence of frost. Variations in topography and distance from the sea produce marked variations in wind speed and direction. The predominant strong wind direction near the coast is southerly. Further inland, the frequency of strong wind decreases and the most common strong wind direction is westerly.

Rainfall is highest on the coast, where the annual average is over 1,200 mm, and on the nearby elevated plateaus with a maximum of 1,440 mm per annum at Pymble. At Parramatta, on the eastern margin on the Cumberland Plain, average annual rainfall drops to around 900 mm, while on the Plain's low-lying central basin, between Pitt Town and Picton, it is less than 800 mm. The driest area is south-west of Campbelltown, where Narellan's annual average, for example, is 670 mm per annum.

In seasonal terms, the Cumberland Plain experiences its wettest period during summer, while near the coast, autumn generally has the most rain. Rainfall is lower in winter. Spring is the driest season across the whole region [1].

Differences in average rainfall exert their effect on species distribution in a number of ways; one of the more important is the length of time which a plant must survive between falls of rain. On the Cumberland Plain sequences of dry days are likely to be longer than on the wetter coast.

Thunderstorms and hail may affect individual plant survival, but there is no evidence that they affect the distribution of particular plant species. Hailstorms are likely to occur, on average, five times every two years at Sydney and three every two years at Liverpool, for example. They have more influence on success of agricultural crops than on native plant distributions. There are about thirty thunderstorms per year, on average, in Sydney, most in late spring and summer.

Local temperatures depend on aspect, altitude and distance from the coast. Mean maximum temperatures increase from less than 26°C along the coast to over 29°C on the Cumberland Plain, while mean minima drop from 7–8°C at Sydney to 2–4°C on the Cumberland Plain. January is the hottest month and July is the coldest. Maximum temperatures of 48.4°C and 46.3°C, and minima of −8.3°C and −2.9°C, have been recorded at Richmond and Parramatta respectively.

Frosts are rare on the coast but common further inland and may influence the distribution of some species. The duration of the frost period increases with increasing distance from the coast and, to a lesser extent, with elevation. On the Cumberland Plain the average frost period may exceed 100 days, mostly between May and September.

2 Plants in Aboriginal Life

from Birra Birra to Dyarrabin

For many thousands of years Aboriginal people lived in the Sydney region. About 6,000 years ago, as the climate stabilised following the end of the last ice age, populations concentrated on the sea coast, north and south of Birra Birra (the lower reaches of Sydney Harbour) and inland, on the floodplain of Dyarrabin (the Nepean–Hawkesbury River). In both places plants provided foods, medicines, tools, weapons, canoes, string bags, and other every day items.

Traditional Aboriginal life at Birra Birra and Dyarrabin has, of course, been destroyed. Smallpox sweeping through the Aboriginal population from 1789, together with European appropriation of the most fertile land, caused a massive disintegration of Aboriginal social structure around Sydney within the first decade of colonial settlement [1]. However, the importance of plants in Aboriginal culture and daily life can be inferred from archaeological evidence, remains of tools, memories of local Aboriginal descendants, the likely distributions of useful plants, and recorded observations of early European settlers.

On the coast plentiful supplies of seafood were supplemented with fruits, nectar, roots and tubers from plants. Extensive sand dunes stretched from Bondi to Botany Bay, and here in the swamps between the dunes, sedges and reeds, *Eleocharis sphacelata, Phragmites australis* and *Triglochin procera* provided edible roots and tubers. In the heath and shrubland on the sandy dunes were sharp-leaved plants with small edible succulent fruits, species of *Astroloma, Leucopogon, Styphelia* and other members of the heath family, Epacridaceae. Shrubby geebungs, *Persoonia* species, Currant Bush, *Leptomeria acida*, and Native Cherry, *Exocarpos cupressiformis* shrubs, and the semi-parasitic Devil's Twine, *Cassytha*, all produced small edible fruits. Ground orchids, *Calochilus paludosus, Cryptostylis erecta, Microtis unifolia, Pterostylis acuminata* and *Thelymitra pauciflora*, for example, and small lilies including *Burchardia umbellata* and *Blandfordia nobilis*, provided edible tubers. Flower spikes of *Banksia* and the *Xanthorrhoea* or Grass Tree were rich in nectar. Botanist George Caley described Aborigines 'collecting the heads of [*Banksia*] flowers and steeping them in water and afterwards drinking it'[2].

Similar plants were found in the shrubby woodland and open-forest of the rocky harbour shores and nearby sandstone country. Here also grew the palm-like cycads or *Macrozamia,* known to the Aborigines as Burrawang; their large seeds provided plenty of carbohydrate, but first needed soaking and cooking to remove toxins. 'They are a kind of nut growing in bunches somewhat like a pine top & are poisonous without being properly prepared the method of doing which we did not learn from them,' wrote Lieutenant William Bradley after a visit to Broken Bay in March 1788 [3]. Preparation lasted several days. The large seeds were broken into chunks, put into a string bag and suspended in a creek or pond for up to eight days, allowing water to leach out the poison. Dried and ground into flour, this was mixed with water to form small cakes, and cooked. Without metal containers for boiling, mixing into a paste or cake was a way foods could be heated.

Trees and vines growing in the sandstone gullies provided fruits; amongst the most abundant were Lillypilly, *Acmena smithii*; Apple Berry, *Billardiera scandens*; Native Grape, *Cissus*; figs, *Ficus*; and native blackberries, *Rubus*. In winter, when berries and fruits were not available, staple foods were starchy rhizomes of Bungwall fern, *Blechnum cartilagineum*, and Bracken, *Pteridium esculentum*, prepared by pounding and cooking. 'Some of the natives were seen, they were all friendly, they seem'd to be very badly off for food not having any fish. . . they were most of them chewing a root much like a fern,' reported William Bradley in May 1788[3].

Fishing parties in canoes would take multi-

In the Eastern Suburbs Banksia Scrub on the sand dunes between Centennial Park and Botany Bay, were many Grass Trees, *Xanthorrhoea resinosa*. Aborigines glued together lengths of flower spike to make spear shafts, using resin from the 'trunk' and base of the plant. The nectar-rich flowers, like those of *Banksia* and *Lambertia*, were steeped in water to make a sweet drink. Several twiners and sharp-leaved heath plants grew small edible fruits. Swollen corky lignotubers of *Banksia serrata* could be hollowed out to make coolamons for carrying water; when leaks developed, *Xanthorrhoea* resin could plug the holes.

Freshwater wetlands provided a wealth of food for Aborigines living on the Nepean–Hawkesbury floodplain. The waterplants *Eleocharis sphacelata* and *Triglochin procera* growing here have starchy edible tubers, while the floating Nardoo, *Marsilea*, is a fern with capsules that can be ground into flour. Beneath Grey Box and Forest Red Gum trees in the surrounding woodlands grew tuberous orchids and lilies, including the Chocolate Lily, *Dichopogon fimbriatus*, and Donkey Orchids, *Diuris*.

pronged fishing spears, and nets and lines made from plant fibres. Their canoes were made from large slabs of bark, cut from River Oak trees, *Casuarina cunninghamiana*. Coastal people travelled inland to cut bark for their canoes from River Oaks along the banks of Dyarrabin, choosing a time soon after heavy rain, when sap was rising, making the bark strong and pliable. William Bradley describes the construction of a boat very different from his own ship, the *Sirius*:

The Canoe is made of the bark taken off a large Tree of the length they want to make the Canoe, which is gathered up at each end and secured by a lashing of strong Vine which runs amongst the underbrush. . . they fix spreaders in the inside, the paddles are about two feet long in shape like a pudding stirrer, these they use one in each hand & go along very fast. In these canoes they will stand up to strike fish at which they seem expert.[3]

Away from the coast, the rolling Cumberland Plain and the fertile alluvial flats provided a different range of foods. Campsites were concentrated along the river, Dyarrabin and its tributary creeks, to take advantage of the abundance and diversity of food plants there[1 4]. Succulent-fruited rainforest plants grew in sheltered alluvial pockets along the river's banks, in deep gullies of the Lower Blue Mountains beyond, and on fertile soils around Kurrajong. East of the river, vines in forest growing on the river flats, lilies, orchids and creepers of the woodlands, and reeds and rushes in the freshwater swamps, all grew edible roots, rhizomes or tubers. 'Yams', tubers of

Thomas Dick asked Aborigines to pose for this photograph near Port Macquarie between 1910 and 1927, one of a series showing how they used to live. Before 1788 there would have been similar scenes in Sydney on mudflats along the Lane Cove River, in Homebush Bay, and near Towra Point. In the foreground, men use stone axes and wedges to prise free a piece of *Casuarina glauca* bark for a shield. Their wooden implements include boomerangs possibly made from *Backhousia myrtifolia* wood, and flattened throwing sticks. The canoes here are made of stringybark, but around Sydney *Casuarina cunninghamiana* bark was used. The man's multipronged fishing spear is made from Grass Tree spikes, while the woman's plant fibre fishing line was fitted with a hook of sharpened shell. (Dick Collection No. V.7784, The Australian Museum)

Dioscorea transversa and other vines, were 'in greatest plenty on the banks of the river; a little way back they are scarce', according to Captain John Hunter in 1793[5]. Vines included *Eustrephus latifolius* and *Geitonoplesium cymosum*; lilies included *Anguillaria, Bulbine bulbosa, Caesia vittata* and *Dichopogon fimbriatus*; amongst the orchids species of *Diuris* were prolific, and the most important swamp plants were probably *Eleocharis sphacelata, Phragmites australis, Triglochin procera,* and *Typha orientalis*.

Aboriginal people mostly camped together in groups of several families, as few as 25 or as many as 120 people sharing a campsite. Seasonal food shortages could be minimised by moving campsites. Plant foods were generally collected by groups of women. Children with them would learn where food plants grew, how to recognise those with edible underground parts, when to expect ripe fruits, and how to spot animal tracks, nests and burrows. Religious obligations to the ancestors and spirits of the Dreaming sometimes dictated the types of food people could collect and eat. For example, nobody could eat a plant or animal that was their own totem. People shared their food with relatives and clan members in a precise order of priority, cementing family and tribal relationships.

Plants provided other everyday items. Circular outgrowths or pieces of eucalypt branch were hollowed out to make coolamons, their carrying dishes. Fibres for making string bags came from the inner bark of the shrubs *Abutilon, Commersonia, Rulingia* and *Hibiscus heterophyllus*, the Native Hibiscus, and roots of *Ficus* trees. Women fished with lines made from the inner bark of the Kurrajong tree, *Brachychiton populneus*. Strips of leaves from Cabbage Palms, *Livistona australis*, giant Gymea Lilies, *Doryanthes excelsa*, and clumps of *Lomandra, Dianella* or *Phragmites*, were used to weave baskets. Axes were made by inserting a sharpened stone axe head into a split length of sapling trunk, and binding with strong twine such as could be made from bark strips of *Pimelea* shrubs[1,6].

Spears, spearthrowers, clubs, axes, shields and throwing sticks were made of different types of wood. *Xanthorrhoea* or Grass Tree flower spikes provided shafts for hunting and fishing spears. Resin from the base of the plant was heated and used to glue on sharp points made of hardwood, stone, bone or shark's tooth. A selection of their implements was described by Surgeon-General John White in August 1788:

In one of their huts, at Broken Bay, which was constructed of bark, and was one of the best I had ever met with, we saw two very well made nets, some fishing lines not inferior to the nets, some spears, a stone hatchet of a very superior make to what they usually have, together with two vehicles for carrying water, one of cork, the other made out of the knot of a large tree hollowed.[7]

The 'cork' water container was probably made from the rootstock of *Banksia serrata*.

Fire was used by Aborigines frequently, its use depending on immediate needs, the time of year, and the type of vegetation around. For making fire, two sticks of different hardness were used. A flat stick was laid on the ground; it needed to be reasonably soft — *Hibiscus heterophyllus* wood was often used, or a piece of *Xanthorrhoea* flower stem, split in half. A harder stick was held upright and twirled very fast between the fingers, to rub away some of the powdery soft wood, heating it to ignition point by friction. The fire-maker kept some tinder handy — fibrous leaf bases of the Cabbage Palm, *Livistona* were ideal — the smoking wood powder was tipped into a handful of the tinder, and whirled very fast at arm's length until it burst into flames. Fire-making was exacting work, and to be avoided where possible, so fire was carried about when collecting food, when moving camp, and on fishing expeditions at night.

Domestic fires were used for cooking, warmth, repelling mosquitoes, sending signals, and special ceremonies. Fire was used to burn larger areas for clearing dense undergrowth, or for driving small animals (possums, echidnas, goannas) out of hollow logs and trees, or larger animals into traps. Fires promote growth and flowering of tuberous plants such as orchids and lilies. These are abundant in the River-flat Forests and in the Cumberland Plain Woodlands and it is likely that Aborigines would have burnt areas of these habitats quite frequently. Inevitably this pattern would have been overlain by the effects of occasional hot wildfires started by lightning or escaped from intentional burns. The resulting patchy mosaic of burnt areas of different sizes, different ages of plant regrowth since last fire, and possibly different types of regrowth depending on fire intensity, provided a diversity of habitats for plants and animals. On the sandstone lands, fires would have been used less often, and probably directed towards localised areas with particular food plants. Fires are known to stimulate the flowering and seed set of some species of *Macrozamia*[8] and *Xanthorrhoea*[9]; the synchronisation of fruit set in this way by fire would have made food collecting more efficient. Burning of some habitats too frequently would have had detrimental effects on other food plants. Heath plants such as *Styphelia* and *Leucopogon*, generally killed by burning, may take 6–8 years after fire to produce good crops of fleshy fruits. Similarly, mesic plants in sandstone gullies would have been killed by fire and take a similar time to re-establish and bear fruit.

Such frequent use of fire seemed strange to Europeans and aroused frequent comments but for Aborigines in Australia it was a part of their way of life. Here, over millions of years, plant species had evolved characteristics that enabled them to survive in naturally fire-prone environments. Different plant communities in the Sydney district would have been susceptible to different fire intensities and rates of spread, as a result of their structure and situation; some, such as saltmarsh and mangroves, would probably never have been burnt. Aborigines worked within this framework to modify plant growth to their advantage.

Aboriginal people were not farmers in the sense of clearing land and planting crops and it is difficult to tell whether they manipulated plant productivity in the Sydney area by means other than fire. William Bradley, participant in several exploratory trips from 1788 onwards, reported 'we never saw the smallest appearance of cultivated land'[3]. Although the distributions of native plants we see today seem to be readily explainable in terms of environmental factors — soil fertility, moisture availability — rather than the result of past plantings or manipulation of conditions, the fertile soils on the alluvial flats, which would have been most responsive to such techniques, were cleared for agriculture by the 1830s, so we cannot find any evidence one way or the other. However, it is known that Aboriginal use of fire maintained a locally diverse environment, unlike the subsequent European-style agriculture that favoured just a few species.

3 Sydney's Vegetation Types

Sydney's bushland vegetation contains different plant communities, each a group of species, growing together in a similar habitat and recurring more or less consistently; each has a particular structure, for example forest, woodland or heath. The structural classification used here, based on the height of the tallest stratum and its canopy cover, has been used widely in ecological studies in Australia[1]. Although only a few large-growing species, commonly trees, may determine the structure, plant communities include many types of plants, from the obvious trees, shrubs and climbers to the inconspicuous herbs, grasses and ferns. We generally only list the dominant species, mostly the trees or shrubs, and probably only five or six species out of perhaps 100 to 150 that may be found at any site. This enables vegetation to be classified or labelled. But it is the rich variety of understorey species which gives many of Sydney's plant communities their beauty and individuality. This diversity is the outcome of ongoing dynamic processes and interactions within plant communities. The occurrence and distribution of uncommon species may provide intriguing clues to Sydney's past history of climatic changes and fluctuating sea levels.

To understand more about a particular site and to appreciate the richness of Sydney's plant life, you will want to identify plant species. Plant identification requires special information and you will probably need to consult plant identification keys and pictorial plant books (see References[2–12]). Lists of plants for specific areas are available from organisations concerned with bushland management and conservation such as local councils, local conservation groups, the National Parks and Wildlife Service, the National Trust and the Royal Botanic Gardens.

Don't assume there will always be information available on your bushland. Despite increasing public awareness, the resources for scientific research are limited and the amount that has been carried out in our bushland is small. The plants and animals in many bushland parks have not been recorded, nor have good management plans been prepared and implemented to ensure their long-term conservation. Concerned local residents have either encouraged local government authorities to prepare such plans, or have carried out inventories and prepared recommendations themselves. To understand natural environments, one of the best ways to learn is by observing the bushland near you.

Sydney's bushland in 1788

Traveller Peter Cunningham described the main features of Sydney's vegetation in 1827:

> In Cumberland, the land immediately bordering upon the coast is of a light, barren, sandy nature, thinly be sprinkled with stunted bushes; while from ten to fifteen miles interiorly, it consists of a poor clayey or ironstone soil, thickly covered with our usual evergreen forest timber and underwood. Beyond this commences a fine timbered country, perfectly clear of bush, through which you might, generally speaking, drive a gig in all directions, without any impediment in the shape of rocks, scrubs or close forest. This description of country commences immediately beyond Parramatta on one hand, and Liverpool on the other; stretching in length south easterly obliquely towards the sea, about forty miles and varying in breadth near twenty. The soil upon the immediate banks of the rivers is generally rich flooded alluvial, but in the forests partakes commonly of a poor clayey or ironstone nature, yet bearing tolerable crops, even without manure, at the outset.[13]

Though written nearly 40 years after the first settlement, the original pattern had been little altered and his comments were still pertinent. Here we see references to the 'stunted bushes' of the Sandstone

Fig. 1 Blue Gum High Forest in Dalrymple Hay Nature Reserve at St Ives.

Fig. 2 Extensive clearing of the Blue Gum High Forest is shown in this view of the Government Farm at Castle Hill, by John Lewin in 1806. (Mitchell Library, State Library of New South Wales)

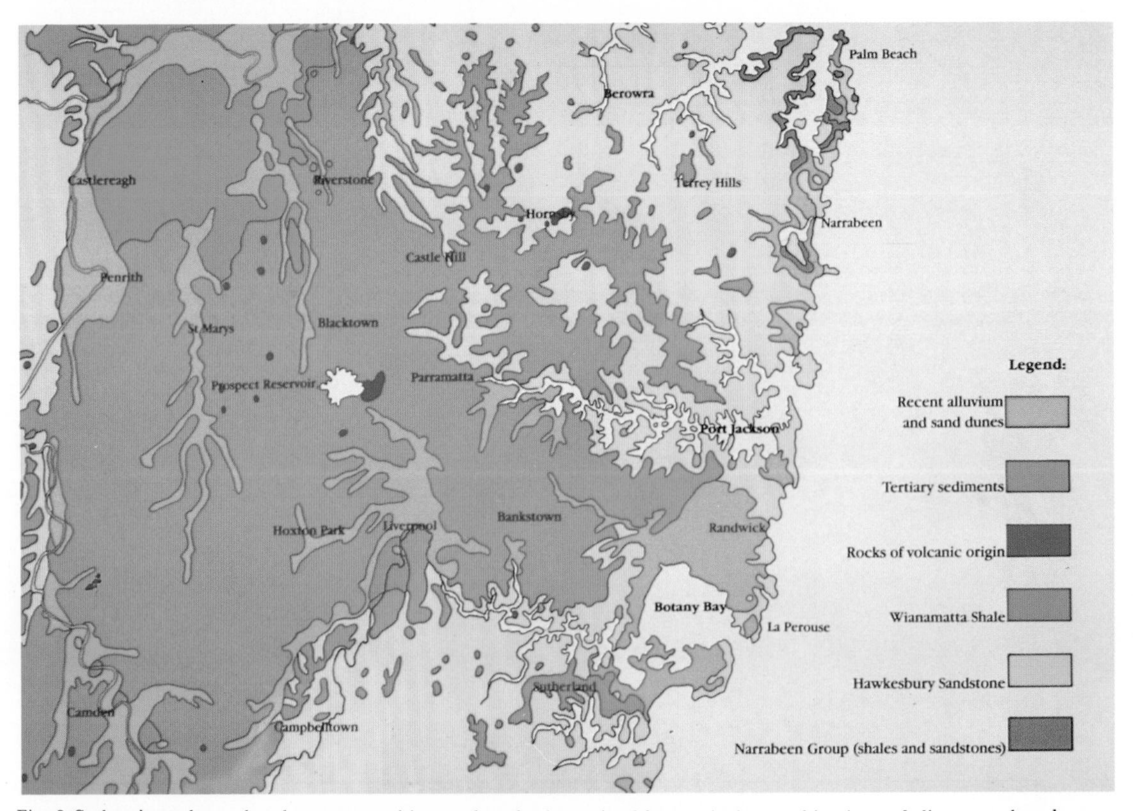

Fig. 3 Sydney's geology: the plant communities are largely determined by particular combinations of climate and geology.

Woodlands and Heaths and the Eastern Suburbs Banksia Scrub, the 'poor clayey or ironstone soil' with Turpentine-Ironbark Forest, and the Cumberland Plain Woodlands' 'fine timbered country, perfectly clear of bush'.

In 1788 there would have been 1,500 native species, of which about 150 were trees and the remainder shrubs, climbers, grasses, sedges, herbs and ferns. There were also mosses, lichens and fungi. Since then the introduction of exotic weed species has added considerably to this number.

The clearing of land for agriculture and the development of the suburbs have obliterated much of the original pattern. In many areas only small bushland remnants or indeed occasional trees are clues to the original vegetation. In other places the vegetation may appear almost undisturbed.

From our survey work we have recognised about 30 different plant communities for the Sydney area, many of which can be subdivided into habitat types depending on local conditions. This complexity is too great to present in this book, so we have grouped the plant communities into eight major vegetation types to cover the major structural and floristic variation.

Blue Gum High Forest

Blue Gum High Forest was found on the high-rainfall areas of Wianamatta Shale soils (receiving more than 1100 mm per year) along the central spine of the North Shore from Crows Nest to Hornsby, and further west on higher land between Castle Hill and Eastwood. This forest was structurally a tall open-forest or wet sclerophyll forest, composed of big trees, probably in places over 40 m in height, and a valuable source of timber last century. Sydney Blue Gum, *Eucalyptus saligna,* and Blackbutt, *Eucalyptus pilularis,* were the main trees, with Blue Gum particularly abundant on the lower slopes and depressions and Blackbutt more prevalent on the ridges. Other tree species were smooth-barked *Angophora costata*; Grey Ironbark, *Eucalyptus paniculata*; White Stringybark, *Eucalyptus globoidea*; Turpentine, *Syncarpia glomulifera*; and Forest Oak, *Allocasuarina torulosa*.

On drier sites the understorey had a layer of shrubs up to 2 m high. Common species were *Dodonaea triquetra, Breynia oblongifolia, Persoonia linearis, Pittosporum revolutum, Leucopogon juniperinus, Platylobium formosum* and *Hibbertia aspera*. Grasses and herbs were frequent, though less conspicuous. On moister sites, particularly in depressions, ferns predominated; particularly *Culcita dubia, Adiantum aethiopicum, Doodia aspera* and *Blechnum cartilagineum*, together with the small trees *Pittosporum undulatum, Glochidion ferdinandi, Clerodendrum tomentosum* and *Polyscias sambucifolia*. Trees with rainforest affinities, including Coachwood, *Ceratopetalum apetalum*, and Lillypilly, *Acmena smithii*, were found along some of the creeks but on Wianamatta Shale soil true rainforest, or 'brush' as it was known in the old days, appears to have occurred only at Brush Farm at Eastwood. Here an unusual combination of rich soils, unusually deep, sheltered gullies, and a high rainfall allowed rainforest species to survive.

Turpentine-Ironbark Forest

On the lower rainfall Wianamatta Shale soils of the inner western suburbs and on the north side from Ryde to Glenorie, the Blue Gum High Forest, requiring good rainfall and deep clay soils, gave way to Turpentine-Ironbark Forest. This was not as tall as the High Forest and would probably have had trees 20 to 30 m high, forming an open-forest structure. There would have been trees of Turpentine, *Syncarpia glomulifera*; White Stringybark, *Eucalyptus globoidea*; Red Mahogany, *Eucalyptus resinifera*; and Grey Ironbark, *Eucalyptus paniculata*.

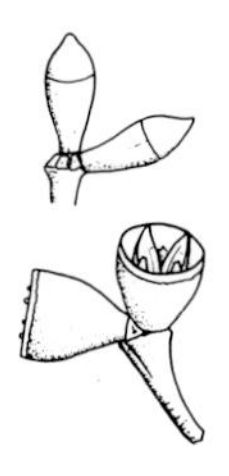

Sydney Blue Gum
Eucalyptus saligna

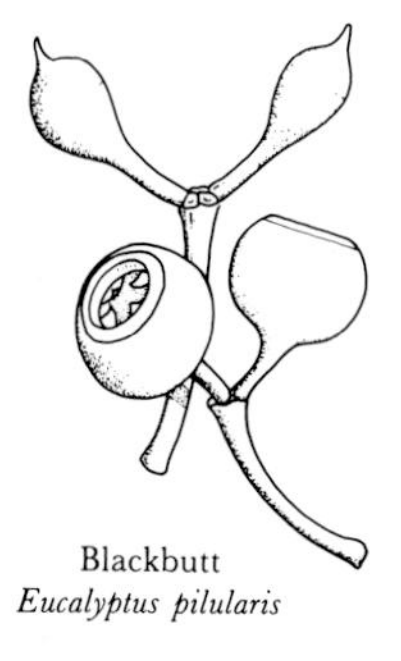

Blackbutt
Eucalyptus pilularis

Common understorey species included *Acacia falcata, Acacia parramattensis, Dodonaea triquetra, Pittosporum undulatum* and *Polyscias sambucifolia*. The change from High Forest to Turpentine-Ironbark Forest would have seen a gradual blending of species as the higher rainfall species dropped out, to be replaced by those more successful on drier sites. Indeed, pockets of High Forest probably occurred in locally better sites. For example, patches of Blackbutt grew in Ashfield and Blue Gum in Burwood.

The Turpentine-Ironbark Forest extended from Glebe and Newtown westward to Auburn. Early descriptions of these areas indicate an abundance of 'heavy timber and brush' (i.e., dense understorey of shrubs). Between Annandale and Ashfield, 'On each side of the [Parramatta] road is a post and rail fence, while the land is thickly covered with heavy timber and brush, the soil being usually a poor shallow reddish or ironstone clay, the contemplation whereof presents but little pleasure to the agriculturalist,' wrote Cunningham in 1827[13].

We can get a very good idea of the understorey species from Mrs Charles Meredith's lively account of her 'bush' at Homebush in the 1840s:

> Many very pretty native flowers and shrubs adorned our 'bush', or rather forest, and the graceful native indigo [*Hardenbergia violacea*] crept up many bushes and fences, sometimes totally hiding them with its elegant draperies. Another handsome climber of the same family (*Kennedia* [*rubicunda*]) has rich crimson flowers, very long in the part called the keel, with bright yellow stamens protruding from its point. This species climbs to a height of twenty or thirty feet, and the dark leaves and drooping flowers hang down in elegant pendulous wreaths. But the most beautiful climbing plant I have yet seen in Australia, I know not the name of except that of *Bignonia australis*, which it probably is [*Pandorea pandorana*]. The leaves resemble those of jessamine in form, but are much larger, and of a rich glossy green; the flowers fox-glove shaped, in long axillary sprays, their colour being a delicate cream-colour, beautifully variegated within by bright purple markings. I only found one plant of it, in a (comparatively) cool moist thicket in our Homebush wood.
>
> Small shrubs with yellow and orange papilionaceous blossoms abounded everywhere, some clinging to the ground like mosses [possibly *Oxylobium scandens*], and others, with every variety of soft and hard, smooth and prickly leaves that can be imagined, growing into tall shrubs [*Dillwynia juniperina, Daviesia ulicifolia, Platylobium formosum*], all very pretty, but with so strong a family likeness that I grew fastidious among them and rarely gathered more than two or three. A small scentless violet [*Viola hederacea*] and a bright little yellow sorrel (which is an excellent salad-herb) [*Oxalis corniculata* group] made some few patches of the dry earth gay with their blue and golden blossoms, and the ground convolvulus [*Polymeria calycina* or *Convolvulus erubescens*] and southern harebell [*Wahlenbergia*] seldom failed to greet me in our rambles. Various kinds of epacris also abounded, with delicate wax-like pink and white flowers [possibly *Lissanthe strigosa* and *Leucopogon juniperinus*].[14]

Although no longer found in Homebush, most of these species can still be found in some nearby bush remnants at Concord or Silverwater.

West of Auburn, around Bankstown, Regents Park and towards Fairfield and Parramatta, the clay soils often have conspicuous ironstone gravels. The rainfall is lower and changes in the vegetation are evident. Drier country trees appear — the Grey Ironbark is replaced by Broad-leaved Ironbark, *Eucalyptus fibrosa*; Turpentine becomes a less common low shrubby tree; and Grey Box, *Eucalyptus moluccana*, and Woollybutt, *Eucalyptus longifolia*, appear, the Grey Box often being particularly common. The understorey ranges from dense scrub up to 3 m high, to open and grassy with scattered shrubs. Dense scrub is found along water courses, the characteristic species being paperbarks; *Melaleuca decora* is the most common in depressions and on creek flats, *Melaleuca styphelioides* is less common but favours creek channels. Smaller shrub species along creeks include

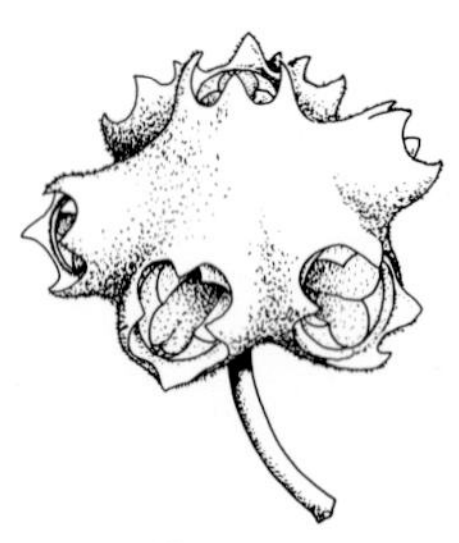

Turpentine
Syncarpia glomulifera

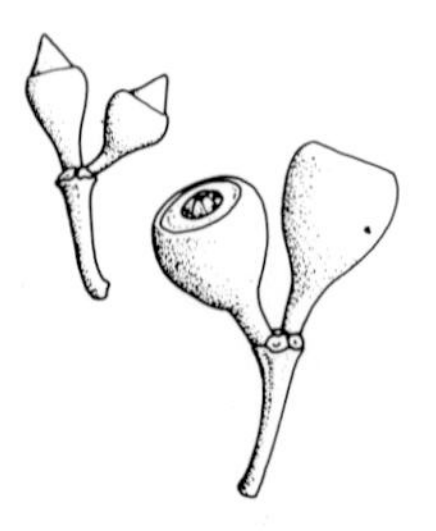

Grey Ironbark
Eucalyptus paniculata

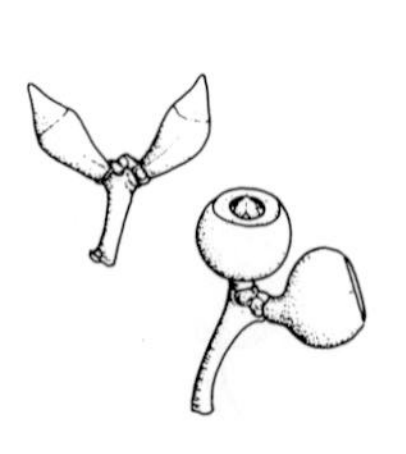

White Stringybark
Eucalyptus globoidea

Red Mahogany
Eucalyptus resinifera

Rapanea variabilis and *Breynia oblongifolia*. On dry gravelly rises shrubs also predominate. Of the larger shrubs *Bursaria spinosa, Melaleuca nodosa* and *Acacia decurrens* are most common. Smaller shrubs include *Lissanthe strigosa, Daviesia ulicifolia, Dillwynia juniperina, Callistemon pinifolius, Acacia pubescens* and *Dodonaea triquetra*. Ground plants, grasses particularly *Themeda australis* and *Aristida vagans*, sedges such as *Lepidosperma laterale*, and herbs such as *Vernonia cinerea, Pratia purpurascens, Hardenbergia violacea, Lomandra longifolia* and *Oxalis corniculata* are interspersed with the shrubs or are more conspicuous on the sides of gravelly ridges.

North of the Parramatta River between Ryde and Glenorie, Turpentine–Ironbark Forest was found on the lower rainfall shale soils, particularly where the shale has weathered to a thin capping over the Hawkesbury Sandstone.

Cumberland Plain Woodlands

West of Parramatta begins the gently undulating Cumberland Plain. It stretches south to Campbelltown and Camden, northwards to Richmond and Windsor, and west to the Nepean-Hawkesbury River. This is the driest part of Sydney; most of the plain receives less than 800 mm of rainfall per annum. Soils are deep clays developed from the shales of the Wianamatta Group, from the Bringelly Shale on the plain itself, and from the Ashfield Shale around the margins. Sandstone strata in the Bringelly Shale outcrop in cliffs on Razorback Range.

In April 1788, seeking better agricultural lands than those at Sydney Cove, Governor Phillip explored country to the west of Parramatta and typical of the Cumberland Plain landscape. He reported:

The country through which they travelled was singularly fine, level, or rising in small hills of a very pleasing and picturesque appearance. The soil excellent, except in a few small spots where it was stony. The trees growing at a distance of from 20 to 40 feet [6–12 m] from each other, and in general entirely free from brushwood, which was confined to the stony and barren spots.[15]

An open, easily penetrable vegetation, free of shrubs, 'brushwood', and presumably with an even grassy groundcover is the impression Phillip gives us of the original vegetation. The distances apart indicate there were about 70 to 200 trees per hectare. Tree densities in remnant stands today are generally much higher, but have a high proportion of younger trees as they are often regrowing after disturbance. The 'stony and barren spots' with 'brushwood' are probably Hawkesbury Sandstone gullies where a more shrubby vegetation would be expected.

Thirty years later James Atkinson describes that 'one immense tract of forest land extends, with little interruption, from below Windsor, on the Hawkesbury, to Appin, a distance of 50 miles'. He had previously indicated that 'forest means land more or less furnished with timber trees, and invariably covered with grass underneath, and destitute of underwood'. Continuing his description of the Cumberland Plain, Atkinson writes: 'The whole of this tract, and indeed all the forest in this county, was thick forest land, covered with very heavy timber, chiefly iron and stringy bark, box, blue and other gums, and mahogany'. He indicates that box and ironbark trees in particular abound in the forest lands[16].

Today the most common and widespread tree species of the Cumberland Plain are the Grey Box, *Eucalyptus moluccana,* and the Forest Red Gum, *Eucalyptus tereticornis*, and these predominated in the woodlands 200 years ago. The Grey Box tends to prefer rises, and the Forest Red Gum the lower hill slopes and depressions. On hilly country these may be accompanied by ironbarks, commonly Narrow-leaved Ironbark, *Eucalyptus crebra*, or perhaps Broad-

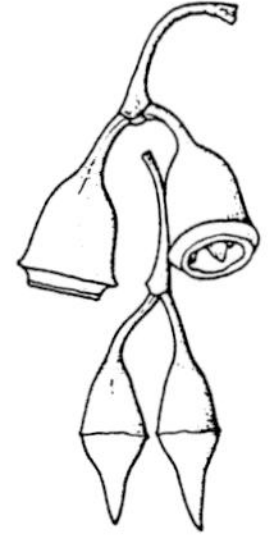

Woollybutt
Eucalyptus longifolia

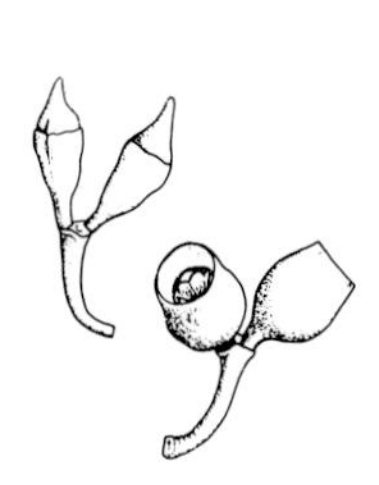

Grey Box
Eucalyptus moluccana

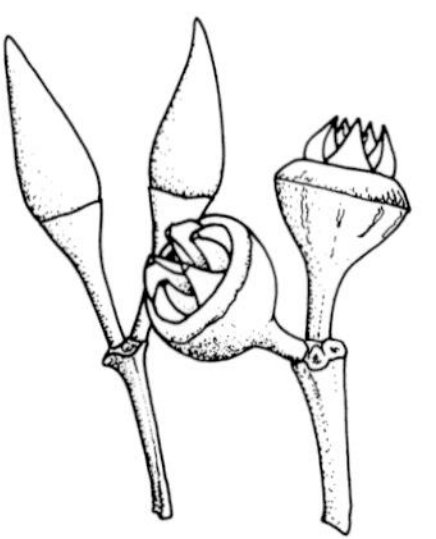

Forest Red Gum
Eucalyptus tereticornis

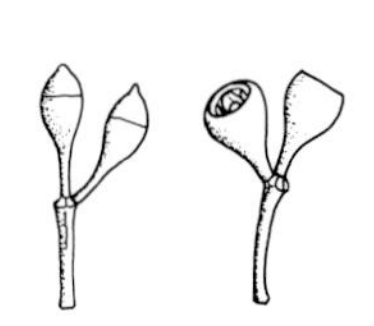

Narrow-leaved Ironbark
Eucalyptus crebra

leaved Ironbark, *Eucalyptus fibrosa*, though the latter often indicates the occurrence of Tertiary ironstone gravels and clays. Stringybark, *Eucalyptus eugenioides*, and Woollybutt, *Eucalyptus longifolia*, occur sporadically, though the main natural occurrence of Woollybutt is further east around Bankstown. Near creeks or on poorly drained sites, Cabbage Gum, *Eucalyptus amplifolia*, Blue Box, *Eucalyptus bauerana*, Bosisto's Box, *Eucalyptus bosistoana*, and Broad-leaved Apple, *Angophora subvelutina* may be found. Such sites may also have groves of Swamp Oak, *Casuarina glauca*, or paperbark, *Melaleuca decora*.

The present understorey may be shrubby or grassy, depending on past disturbance or grazing treatments. The most common shrub species is *Bursaria spinosa*, which may be found in dense clumps or as scattered individuals. Less common shrubs are *Dillwynia juniperina, Daviesia ulicifolia* and *Indigofera australis*. Where the soils have been undisturbed, native perennial grasses still occur, particularly *Themeda australis, Eragrostis leptostachya, Aristida vagans* and *Aristida ramosa,* and the herbs *Brunoniella australis, Lomandra filiformis, Dianella laevis* and fern *Cheilanthes sieberi*. Where the soils have been ploughed or fertilised, exotic grasses such as *Paspalum dilatatum* predominate.

The relative abundance of shrubs and grasses at the time of settlement is now impossible to determine. The early writers describe a general lack of underwood, but with localised patches of shrubs. On visiting a farm at Liverpool in 1817, the botanist Allan Cunningham (not related to Peter Cunningham) wrote: 'Like other farms in the neighbourhood it is overrun with the *Bursaria spinosa* now in fruit'[17]. *Bursaria* may have increased after settlement as a result of cultivating, changes in grazing, fire frequency, or a combination of these events.

Variations in floristic composition would have been caused by soil and drainage conditions. John Macarthur complained to Governor King in 1805 that land west of Liverpool was unsafe to feed sheep: 'Almost the whole is of that wet kind which has been found so fatal to Sheep, or is covered with Scrubby brush Wood'[18]. The 'wet kind' of land, that probably caused sheep losses through liver fluke or footrot, would probably have been poorly drained land with stands of *Casuarina glauca*. 'Scrubby brush wood' may refer to dense stands of *Melaleuca decora*. These are still common west of Liverpool. By comparison, woodlands around Camden were 'chiefly of undulating, thinly-wooded hills covered with a sward of fine dry native pasture' and 'dry, firm and... in every respect so well adapted for a Sheep Pasture'[18].

Many native species resprout vigorously after fire. Here *Bursaria spinosa*, a common Cumberland Plain Woodland species, is regrowing from the stem base.

Vegetation associated with the more rugged escarpments of Razorback Range and its spurs, particularly Donalds Range, often had a vine thicket understorey which in places approached dry rainforest. Remnants are found on outcrops of the sandstone strata of the Bringelly Shale, along gullies and cliff lines on sheltered north-east to south-east aspects. Small tree species with rainforest affinities found on Razorback include the well known Red Cedar, *Toona australis*, together with *Notelaea longifolia, Guioa semiglauca, Claoxylon australe, Alectryon subcinereus, Rapanea variabilis* and *Myoporum montanum*. Lianes and scramblers are plentiful, common species including *Clematis glycinoides, Aphanopetalum resinosum, Cayratia clematidea* and *Stephania japonica*. One vine, *Cissus opaca*, was not previously known to occur south of Port Stephens until collected on Razorback Range in 1968. There appears to have been a sparse understorey with banks of fern such as *Pellaea falcata* and *Adiantum aethiopicum*. Unfortunately this vegetation has been considerably altered by extensive removal of much of the tree canopy of Narrow-leaved Ironbark, *Eucalyptus crebra*, and Coastal Grey Box,

Eucalyptus quadrangulata, and the invasion of the introduced African Olive, *Olea europaea* subspecies *africana*, which forms a dense scrub preventing native shrub and herb regeneration.

Castlereagh Woodlands

In the north-western corner of the Cumberland Plain, between Penrith and Windsor, quite different vegetation from the Grey Box–Forest Red Gum woodlands may be found. Here around Castlereagh and Londonderry are soils developed on alluvial sediments of Tertiary age; clays, silts, sands and gravels deposited millions of years ago by an ancestral Nepean-Hawkesbury River system. The country is flat with shallow meandering drainage lines and the soils are sands or clays, often with a conspicuous component of ironstone gravels or larger river stones. They are all very poor in nutrients, and unlike the more fertile Wianamatta Shale soils, have not been used for agriculture to any extent. Large areas still remain reasonably undisturbed.

'The country generally is heavily timbered with the iron-bark tree, a little distance from the water-courses; with here and there tea-tree brush'[19] described travellers Samuel Mossman and Thomas Banister in 1853. Ironbarks, mainly Broad-leaved, *Eucalyptus fibrosa*, but also some Narrow-leaved, *Eucalyptus crebra,* and Mugga, *Eucalyptus sideroxylon*, are characteristic of the gravelly clay soils today. Where the soil is more sandy these trees are replaced by Scribbly Gum, *Eucalyptus sclerophylla*, and Narrow-leaved Apple, *Angophora bakeri*. Along watercourses and poorly drained depressions are paperbark, *Melaleuca decora*, and Drooping Red Gum, *Eucalyptus parramattensis*. There are also changes in the shrub and ground components. The ironbark-dominated forests have shrubs of *Melaleuca nodosa* and *Melaleuca decora, Daviesia ulicifolia, Pultenaea villosa* and *Dillwynia tenuifolia*. With the Scribbly Gums are *Ricinocarpos pinifolius, Pimelea linifolia, Banksia spinulosa, Hakea dactyloides* and *Leptospermum attenuatum*. The poorly drained sites have more grasses and herbs and include some now uncommon species such as *Elatine gratioloides, Gratiola pedunculata, Haloragis heterophylla* and *Isotoma fluviatilis*.

River-flat Forests

The Nepean-Hawkesbury River floodplain, particularly downstream from Windsor, is characterised by high alluvial levee banks separating depressions or 'back-swamps' with Freshwater Wetland from the main river. Tall open-forest with eucalypts over 30 m in height was found on these levee banks. 'Most of the alluvial lands were originally forest; the timber was large, principally blue and flooded gum, with an abundance of the tree known in the Colony by the appellation of the apple tree, which is of little value,' wrote James Atkinson in 1826[16]. His 'blue and flooded gum' refer variously to trees of Sydney Blue Gum, *Eucalyptus saligna*; Deane's Gum, *Eucalyptus deanei*; Forest Red Gum, *Eucalyptus tereticornis*; and Cabbage Gum, *Eucalyptus amplifolia*, all of which would have been found along the Hawkesbury River below Penrith. The 'apple tree' was *Angophora subvelutina*, the Broad-leaved Apple, found on the main floodplain, or the rough-barked *Angophora floribunda*, also found on the flats along the smaller creeks. River Oaks, *Casuarina cunnninghamiana*, and occasional Water Gums, *Tristaniopsis laurina*, were found along the river edge downstream to Wisemans Ferry.

The understorey of the floodplain forests would probably have had a discontinuous shrub layer of *Bursaria spinosa*, with grasses including *Eragrostis brownii* and *Aristida vagans* on well drained sites. Along the levee banks the understorey appears to have been particularly dense with shrubs and vines, and made travelling difficult, as John Hunter relates:

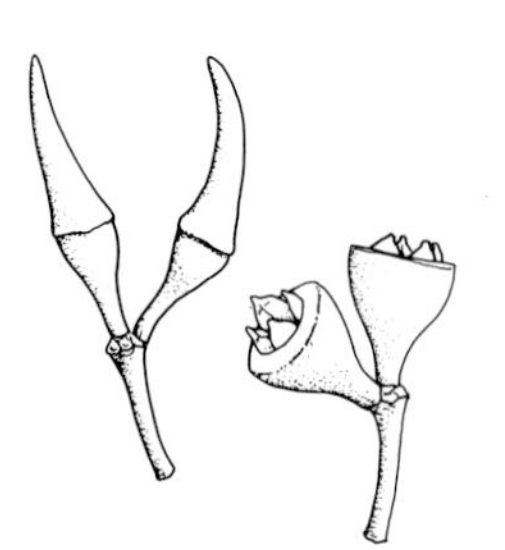

Broad-leaved Ironbark
Eucalyptus fibrosa

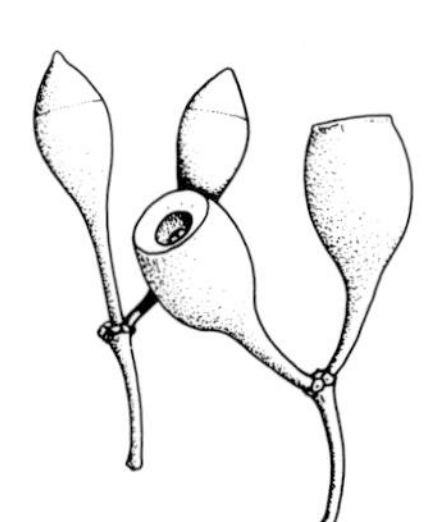

Mugga Ironbark
Eucalyptus sideroxylon

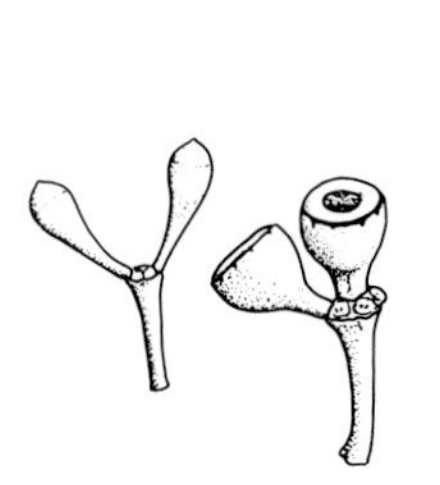

Scribbly Gum
Eucalyptus sclerophylla

Rough-barked Apple
Angophora floribunda

The person who was charged with counting his paces, and setting objects to which they directed their march, had hitherto gone first; but the long sedge [*Gahnia, Lomandra longifolia*], the dead branches which had fallen from the trees, the nettles [*Urtica incisa*], and a weed resembling ivy which entangled the feet [*Smilax australis*?], made walking on, or near the banks of the river very fatiguing.[20]

On sites sheltered by the steep-sided sandstone cliffs, particularly downstream from Sackville, there were pockets of low dense forest with small trees and shrubs, many of 'rainforest type' species. They included small trees of *Acmena smithii, Trema aspera, Ficus coronata, Duboisia myoporoides, Backhousia myrtifolia, Tristaniopsis laurina* and Red Cedar, *Toona australis*. Vines *Eustrephus latifolius, Geitonoplesium cymosum* and *Pandorea pandorana* twisted among the shrubs, while on the ground, banks of ferns, commonly *Doodia aspera* and *Adiantum aethiopicum*, grew in the restricted light in the moist litter layer. Where moisture and shelter were less favourable, a more open shrub layer of *Leptospermum flavescens* and bipinnate wattles such as *Acacia parramattensis, Acacia decurrens* and *Acacia filicifolia* was present. Ground cover was *Pteridium esculentum* and native grasses. *Lomandra longifolia* was frequent, particularly on the river banks.

Above the junction with the Grose, the river is known as the Nepean, and on its floodplain would have been mainly Cabbage Gum, *Eucalyptus amplifolia*, and Broad-leaved Apple, *Angophora subvelutina*. Rough-barked Apple, *Angophora floribunda*, appears to have replaced *Angophora subvelutina* along the smaller, slower flowing Cumberland Plain creeks such as South, Kemps and Badgerys, though these two *Angophora* species also interbreed. Ribbon Gum, *Eucalyptus viminalis*, and Bangalay/Blue Gum, *Eucalyptus botryoides/saligna* intergrades, were found along the Nepean between Menangle and Camden, and River Peppermint, *Eucalyptus elata*, as far downstream as Wallacia. The now rare Camden White Gum, *Eucalyptus benthamii*, although now restricted to Cobbitty-Wallacia, occurred sporadically along the river.

Where the river passes through sandstone gorges south of Penrith the alluvial vegetation is very restricted in area. A fringe of *Casuarina cunninghamiana* and *Angophora subvelutina* is common, with *Eucalyptus deanei* on the lower slopes.

The Nepean's riverbank scrub would have included small trees of *Acacia binervia, Tristaniopsis laurina, Backhousia myrtifolia, Acmena smithii, Acacia floribunda, Melia azedarach* and *Trema aspera*. Smaller shrubs would have included *Hymenanthera dentata, Phyllanthus* species, *Goodenia ovata,* and *Duboisia myoporoides,* with ground cover plants the grasses *Microlaena stipoides* and *Stipa verticillata*, herbs *Geranium homeanum* and *Pratia purpurascens,* and ferns *Adiantum aethiopicum* and *Pellaea falcata*.

Other more localised tree species included Blue Box, *Eucalyptus bauerana,* and Bosisto's Box, *Eucalyptus bosistoana*. Of *Eucalyptus bauerana,* the naturalist William Woolls wrote in 1880: 'I have found a tree of this species occasionally amongst forest-trees in the neighbourhood of Liverpool and Richmond but its proper habitat is near the banks of rivers and creeks. It does not occur as far as I have been able to ascertain, near Sydney or Parramatta'[21]. *Eucalyptus bauerana* is still common near Picton and along the Georges River near Milperra. It is a tree of slow growth and there are some large old specimens along the Nepean River at 'Camden Park', though some are threatened by sand-mining operations. *Eucalyptus bosistoana* used to occur at Milperra on the Georges River.

On the floodplains of sluggish or intermittent tributary creeks, Swamp Oak, *Casuarina glauca*, frequently formed dense stands. Clumps of small trees are often the result of its root-suckering nature. This species is characteristic of saline estuarine situations[2], and its distribution appears related to the presence of the saline groundwater found below

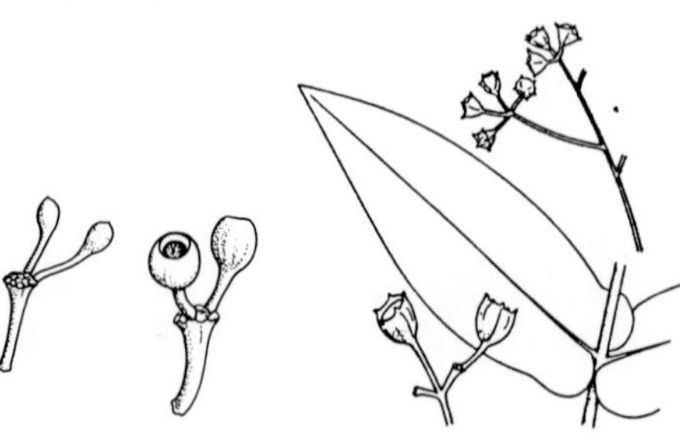

River Peppermint
Eucalyptus elata

Broad-leaved Apple
Angophora subvelutina

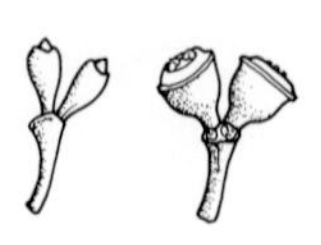

Camden White Gum
Eucalyptus benthamii

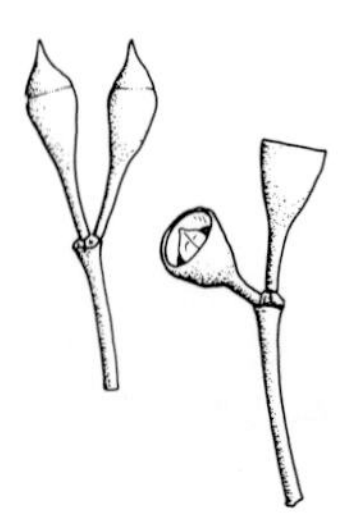

Blue Box
Eucalyptus bauerana

much of the Wianamatta Shale country, particularly below the low-lying central Cumberland Plain[22].

The River-flat Forests occupied the most fertile agricultural soil in the Sydney area, land first farmed in the 1790s. 'The greater part of the alluvial lands upon the Hawkesbury and Nepean have been cleared, and are under cultivation,' reported James Atkinson in 1826[16]. A few scattered old trees surrounded by improved pasture are generally all that remain.

Sandstone Heaths, Woodlands and Forests

In the popular perception, Sydney bushland evokes images of the Hawkesbury Sandstone landscape and its vegetation. 'The general face of the country. . . diversified with gentle ascents, and little winding vallies, covered by the most part with large spreading trees, which afford a succession of leaves in all seasons,'[23] impressed the European settlers on their arrival in 1788. The 'large spreading trees', however, disguised the poverty of the sandstone soils and the settlers soon reported 'the soil in and about the settlement seems to be very indifferent and unproductive'[24]. The shallow sandy soils were very low in phosphorus, a nutrient important for plant growth, particularly for agricultural plants. The native Hawkesbury Sandstone flora had evolved a very rich and distinctive assemblage of species that could thrive on the poor soils; characteristically these are sclerophyllous plants with tough, often small or spiky leaves. The rugged sandstone and its poor agricultural soils were avoided by the settlers seeking farming land, and today it is the open-forests, woodlands and heaths of the sandstone that make up the largest areas of remaining natural vegetation around Sydney.

Vegetation patterns on the sandstone landscapes respond strongly to the variety of local habitats. Topography affects available soil moisture; steeper slopes tend to be better drained than gentler ones, and slopes facing north and west receive more sunlight, drying out faster than those facing south and east. South-facing slopes are generally steeper than north-facing ones. Deeper soils accumulate downslope, less exposed to the drying effects of the sun, providing moister and often more fertile conditions for plants. Lenses of shale, interbedded among the sandstone layers, weather to pockets of clay-rich soils, with higher fertility and better water-holding capacity than the sandstone soils. Shale strata may also concentrate water, providing locally wetter conditions in springs or soaks.

Vegetation structure ranges from open-forests in gullies, on lower slopes and other sheltered sites, to woodland on ridges and more exposed slopes, and heath on sites with very shallow or poorly drained soils. Shrub and ground cover plants largely contribute to a rich diversity of species. In many areas it is common to find 60–80 different plant species growing together in an area half the size of an average house block. Despite considerable local variation, groupings of plant communities related to habitats can be recognised.

Heath and scrub on shallow or poorly drained soils

'The barren scrubs almost every where border the sea coast, and extend to varying distances inland,' recorded James Atkinson in 1826.

> The soil in these scrubs is either sandstone rock or sterile sand or gravel, covered, however, with a profusion of beautiful shrubs and bushes, producing the most elegant flowers, and affording a constant succession throughout the whole year, but most abundant in winter and spring; the shrubs and plants growing in these places furnish the Colonists with materials for brooms, but produce little else that can be converted to any useful purpose.[16]

Coastal heath was the main vegetation on coastal headlands from Palm Beach to Royal National Park,

Dwarf Apple
Angophora hispida

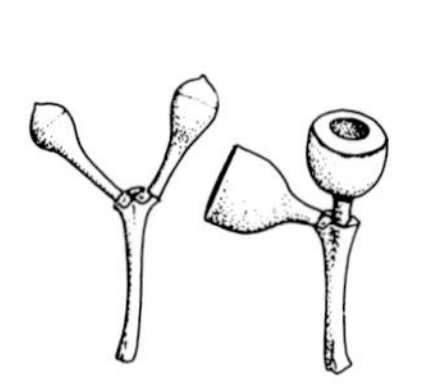
Scribbly Gum
Eucalyptus racemosa

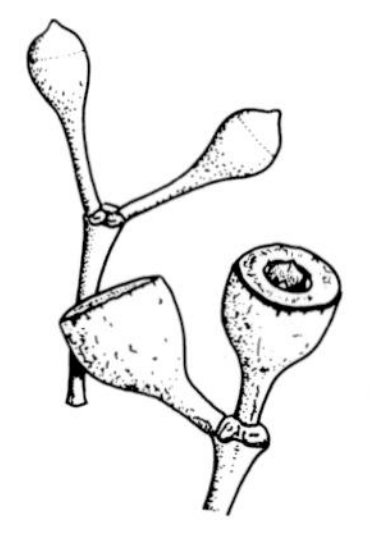
Scribbly Gum
Eucalyptus haemastoma

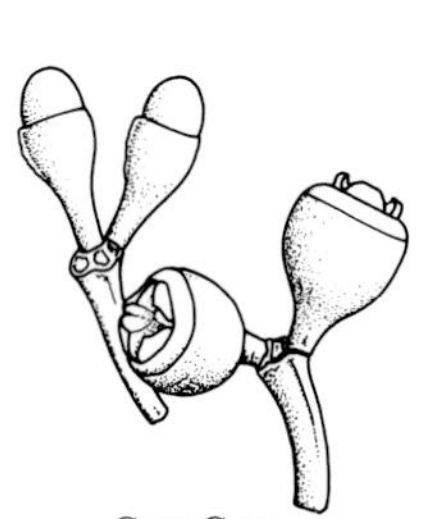
Grey Gum
Eucalyptus punctata

while extensive areas of heath and scrub were found around Middle Harbour, Deep Creek and more sporadically in the Lane Cove valley, Hornsby to Kenthurst, and Sutherland to Holsworthy. Because it occupied ridges, much has been destroyed for houses.

Today, heath and scrub is found on shallow exposed soils or where drainage is impeded by rock shelves or shale lenses. Here *Banksia ericifolia, Angophora hispida, Allocasuarina distyla* and needle-leaved *Hakea teretifolia* grow in dense thickets up to 4 m tall. There may be occasional emergent trees of *Eucalyptus haemastoma, Eucalyptus gummifera,* or *Eucalyptus oblonga.* Where soil is damper, the shrub layer diminishes and ground cover sedges predominate.

Mallee eucalypts — *Eucalyptus luehmanniana, Eucalyptus obtusiflora, Eucalyptus multicaulis* — typically grow in shallow sandy soils on or below ridge lines, and may be associated with seepage zones.

Shallow pockets of soil on sandstone platforms support islands of low shrubs above moss and lichen carpets. *Baeckea brevifolia, Baeckea diosmifolia, Darwinia fascicularis, Calytrix tetragona, Allocasuarina distyla, Leucopogon microphyllus* and *Kunzea capitata* can often be found here.

Woodlands on ridge-tops and exposed upper slopes

Woodland occupied extensive tracts of sandstone country around Sydney, and may be seen today in many bushland parks and reserves. In woodland on ridge-tops and upper hillslopes Scribbly Gums are probably the most characteristic trees. *Eucalyptus haemastoma* is most likely to be found on purely sandy soils, and *Eucalyptus racemosa* where there is a clay influence; the species are often found growing together because of soil patchiness. Red Bloodwood, *Eucalyptus gummifera,* and the Smooth-barked *Angophora costata* are common. Stringybarks are *Eucalyptus oblonga* or *Eucalyptus sparsifolia.* Grey Gum, *Eucalyptus punctata,* occurs where there is some shale or ironstone influence. *Eucalyptus sieberi,* Black Ash or Silver-top Ash, occurs sporadically on ridge-tops between Broken Bay and Sydney, but is more common south of Sutherland. North-west of Sydney, in Arcadia and Maroota for example, *Eucalyptus eximia,* the Yellow Bloodwood, occurs quite extensively.

Woodland understories are generally shrubby; common species include *Banksia serrata, Boronia ledifolia, Dillwynia retorta, Lambertia formosa, Leptospermum attenuatum, Petrophile pulchella* and *Pultenaea elliptica.*

Open-forest on exposed hillsides

Like woodland, open-forest is still common in bushland parks and reserves. On exposed slopes, Sydney Peppermint, *Eucalyptus piperita*, the smooth-barked *Angophora costata*, and occasional Red Bloodwoods, *Eucalyptus gummifera* may be found. *Allocasuarina littoralis*, the Black She-oak, is a frequent small tree, and often forms dense thickets.

Open-forest on sheltered hillsides

On sheltered hillsides, open-forest may also contain Turpentines, *Syncarpia glomulifera,* and small trees of *Elaeocarpus reticulatus*, Blueberry Ash, Christmas Bush, *Ceratopetalum gummiferum,* and *Pittosporum undulatum.* The shrub layer may be dense and diverse, and include softer shrubs such as *Grevillea linearifolia* and *Persoonia pinifolia.* On the forest floor *Billardiera scandens* and *Smilax glyciphylla* twine amongst ferns — Bracken, *Pteridium esculentum,* and the bracken-like *Culcita dubia.*

Creekside scrub and forest

Small permanent creeks with rocky banks are common in sandstone country. Here are small trees and shrubs of *Tristaniopsis laurina, Callicoma serratifolia* and *Lomatia myricoides.* Small shrubs, *Austromyrtus tenuifolia* and *Stenocarpus salignus,* and ferns including *Gleichenia dicarpa, Sticherus flabellatus,* filmy *Hymenophyllum,* and Maidenhair, *Adiantum*

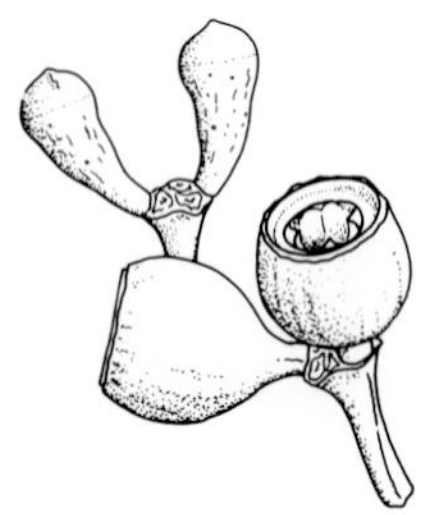

Port Jackson Mallee
Eucalyptus obtusiflora

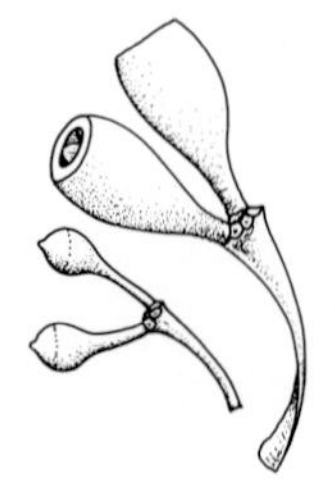

Black Ash Silver-top Ash
Eucalyptus sieberi

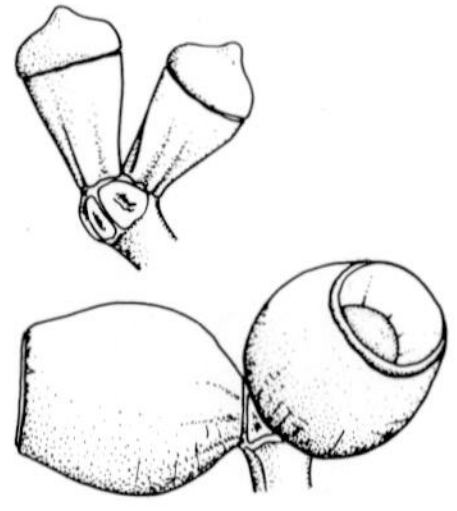

Yellow Bloodwood
Eucalyptus eximia

Smooth-barked Apple
Angophora costata

aethiopicum, grow amongst the sandstone boulders. Along streams receiving stormwater run-off from houses and factories, weeds such as Privet rapidly invade and choke out native plants.

On valley floors where alluvial soils are more fertile and extensive, such as along the Lane Cove River, tall open-forest with Blackbutt, *Eucalyptus pilularis*, Sydney Blue Gum, *Eucalyptus saligna*, Grey Ironbark, *Eucalyptus paniculata,* and Turpentine, *Syncarpia glomulifera*, once occurred. Soft-leaved shrubs, *Polyscias sambucifolia* and *Pultenaea flexilis,* and the vines *Pandorea pandorana, Clematis aristata* and *Hibbertia dentata* grew with the ferns *Pteridium esculentum, Culcita dubia* and *Blechnum cartilagineum*. Where alluvial soil was locally enriched by fertile downwashed clay, small rainforest trees grew under eucalypts up to 30 m tall. Rainforest species included Lillypilly, *Acmena smithii,* Coachwood, *Ceratopetalum apetalum,* Cedar Wattle, *Acacia elata,* and Lawyer Vine, *Smilax australis*.

Vegetation of the Narrabeen Group

Narrabeen Group strata outcrop beneath the Hawkesbury Sandstone in a band of up to three kilometres wide along the coast between Long Reef and Barrenjoey, and on the lower slopes of hillsides around Pittwater, Broken Bay and along the Hawkesbury River as far as Wisemans Ferry. These sedimentary strata are richer in shale than those of the Hawkesbury Sandstone and weather quickly to more fertile soils. The Narrabeen Group is included within the sandstone landscape for convenience because of its limited geographical extent.

There were forests of Spotted Gum, *Eucalyptus maculata,* and Grey Ironbark, *Eucalyptus paniculata,* with localised rainforest patches with Cabbage Palm, *Livistona australis*, on the northern beaches peninsula. Remnants may still be seen at Bilgola and Avalon. Drier forest with Rough-barked Apple, *Angophora floribunda*, as a common species, was found around Pittwater and Broken Bay and along the Hawkesbury River.

Eastern Suburbs Banksia Scrub

In the Eastern Suburbs most of the Hawkesbury Sandstone is covered by Quaternary sands and clays. Extensive wind-blown dune sands cover the central core between Moore Park and Bunnerong, while along the Cooks River and Sheas Creek are estuarine silts and clays. Sclerophyllous heath, scrub and low forest originally grew on the sand dunes. 'Crowded with such exquisite flowers that to me it appeared one continued garden,' enthused Louisa Meredith in 1839[14]. Surviving remnants at La Perouse, Eastlakes and Centennial Park give a glimpse of its original richness. Common shrub species were *Banksia aemula, Banksia serrata, Monotoca elliptica, Eriostemon australasius, Ricinocarpos pinifolius* and *Xanthorrhoea resinosa*. Small soaks and concentrations of organic matter in the sand formed locally wet habitats for *Goodenia stelligera, Callistemon citrinus,* Button Grass, *Gymnoschoenus sphaerocephalus,* and other swamp heath plants. The variety of habitats contributed to the rich assemblage of species in the Banksia Scrub.

Banksia aemula, a north coast species, reaches its southern distributional limit at La Perouse; in studies at Myall Lakes it has been found only on older leached sands, and appears to be restricted to such sands in the Eastern Suburbs. The related and superficially similar *Banksia serrata* appears to favour younger higher nutrient sands or Hawkesbury Sandstone sites. *Angophora costata* grows as a mallee or small tree in similar places. In swales and drainage lines between the sand dunes were freshwater swamps.

On Botany Bay's southern shore, similar vegetation grew on sand dunes between Cronulla and Kurnell, though here *Banksia aemula* does not occur. Most of the vegetation on the older sands here has been lost to sand mining and industrial development.

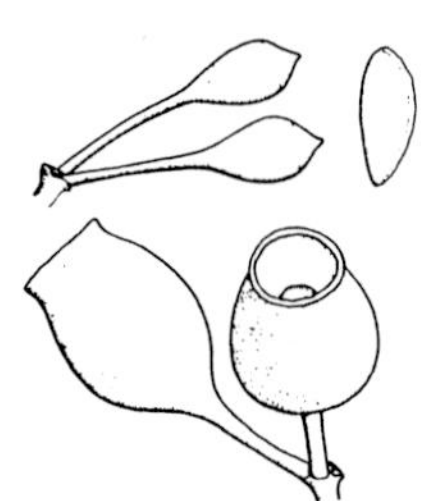

Red Bloodwood
Eucalyptus gummifera

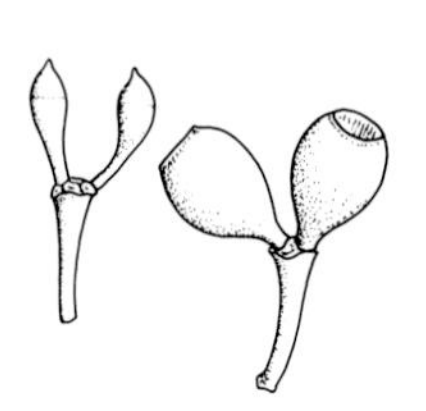

Sydney Peppermint
Eucalyptus piperita

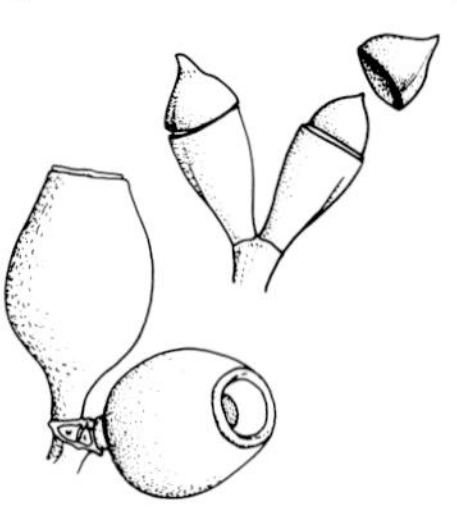

Spotted Gum
Eucalyptus maculata

Younger sand dunes overlie sandstone in Captain Cook's Landing Place Historic Site.

Freshwater and Estuarine Wetlands

Habitats with an oversupply of water, or where the water fails to drain away quickly, are known as wetlands or swamps. Most plants are unable to grow in very wet soils because there is not enough oxygen for their roots. Plants found in these habitats generally have special adaptations that enable them to cope. Wetlands may have either fresh or salt water, or an intermixing of the two.

Freshwater Wetlands — on the floodplains

Freshwater Wetlands were associated with the floodplains of the main rivers, and in particular with the floodplain of the Nepean-Hawkesbury River. Here natural levee banks built up along the river cut off low-lying back-swamps between the levee and the outer edge of the floodplain. 'There are many flat pieces of land in the neighbourhood of the Rivers, that are covered with what is termed *backwater* in time of floods,' wrote James Atkinson in 1826[16]. Such back-swamps had a series of vegetation zones related to the depth of flooding and period of inundation, though grazing and draining activities have obscured these in many remaining swamps.

More or less permanent standing water occupied the most poorly drained, and often central, section of these swamps. Where this was more than 3 or 4 m deep it was usually open water, lacking emergent vegetation but with some free-floating species. Shallower areas of open water had emergent reeds and rushes, particularly the Tall Spike Rush, *Eleocharis sphacelata,* growing up to 1–2 m high above the surface of the water. Other smaller species such as *Triglochin procera, Ludwigia peploides* subspecies *montevidensis* and *Philydrum lanuginosum* were found, with their leaves floating on the water surface, but with roots embedded in the bottom of the pool. Some species were indicative of particular water conditions. For example, *Typha*, Bullrush species, have readily wind-dispersed seeds and quickly colonise new ponds or farm dams, particularly if they have a high nutrient content, while the reed *Phragmites australis*, one of the grass species, inhabits sites where the water may be brackish or even saline, and may be common in the upper reaches of estuaries.

Where the soil was waterlogged or very moist, but standing water was only present after rain, rushes grew, perennial tussock plants of the Juncaceae family. The most common floodplain species is *Juncus usitatus*. With the rushes were quick-growing herbs, commonly *Persicaria,* particularly the annual *Persicaria lapathifolia* or the perennial *Persicaria strigosa*.

Around the *Juncus* zone was paperbark shrubland. Shrubs or small trees of the paperbarks *Melaleuca linariifolia* or sometimes *Melaleuca styphelioides* were once common at the head of the typical Hawkesbury back-swamps but have disappeared as old trees have died without being replaced. Grazing stock have eaten *Melaleuca* seedlings or, by trampling the soil, made establishment of the necessary seedlings difficult. Damming to maintain permanent water also slowly kills these plants which tolerate periodic but not permanent inundation. In most places today only remnants with ageing plants remain.

Close to the paperbark shrubland may have been swamp woodland of Swamp Mahogany, *Eucalyptus robusta*. Almost all of these areas have now been cleared.

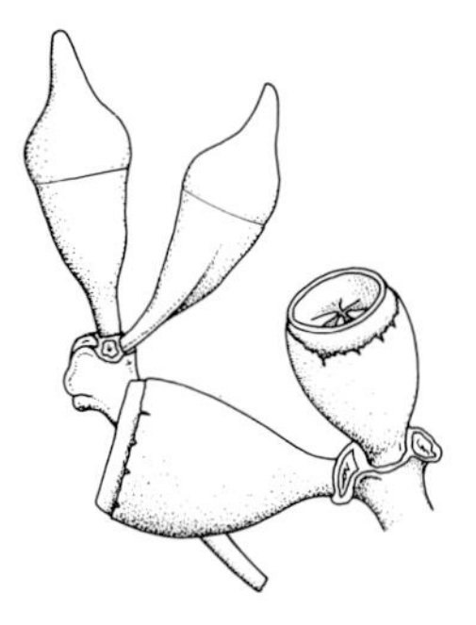

Swamp Mahogany
Eucalyptus robusta

Because of the agricultural value of the floodplain soils, back-swamps have been drained to provide additional pastures. Most remaining swamps have been grazed for nearly 200 years and the most palatable native species have disappeared, to be replaced by a variety of introduced pasture plants. Persistent species like *Juncus usitatus* are not readily eaten by stock. Some swamps still remain along the Colo River, the Macdonald River, for example, the Common at St Albans, and the Hawkesbury River particularly between Sackville and Lower Portland and along Little Cattai Creek. As well as native plants, these provide valuable habitat for waterbirds, and pollen and charcoal in their accumulated sediments may hold clues to changes in Sydney's vegetation since the end of the last ice age.

Freshwater Wetlands — on the coastal sands

Freshwater Wetlands were also found in the swales and drainage lines of the sand dunes of the Eastern Suburbs, particularly in the Lachlan Swamps, today's Centennial Park, and in the Botany Swamps at Eastlakes. Here, around expanses of open water with patches of tall *Eleocharis sphacelata*, were shorter sedges, *Baumea articulata, Baumea rubiginosa* and *Juncus* species. Nutrients here were lower than on the floodplain and shrub species would have also been present, though only a few can be found today. The most conspicuous species is the paperbark *Melaleuca quinquenervia* which may occur as a small or large tree. There is a healthy patch of small trees in the Mill Stream of the Botany Swamps, and a group of larger trees in the Lachlan Swamps at Centennial Park. Similar vegetation was evidently found between Bondi and Rose Bay[25]. Native Broom, *Viminaria juncea*, was also a common component of the freshwater swamps. Arthur Hamilton wrote in 1918: 'In the Botany swamps the *Viminaria* frequently forms large shrubberies which are occasionally devastated by fire'[25]. Other shrubs in these swamps may have been *Callistemon citrinus* and *Callistemon linearis, Leptospermum juniperinum* and *Kunzea ambigua*.

Estuarine Wetlands

Estuarine Wetlands — those wetlands found in the harbours and bays along the coast and influenced most strongly by the salinity of the water — are characterised particularly by mangroves and saltmarsh. Mangrove vegetation was recorded as a characteristic part of the Sydney estuaries in 1788. 'For it is strikingly singular that three such noble harbours as Botany Bay, Port Jackson, and Broken Bay, alike end in the shallows and swamps, filled with mangroves,' wrote Watkin Tench[23]. Lieutenant Bradley confirms this: 'Those Coves above where the ships lay were surrounded by Mangroves and had Mud flats at the bottom, those below had sandy beaches most of them'[24].

Sydney mangroves are largely dominated by one plant species, the Grey Mangrove, *Avicennia marina*, a shrub or small tree up to 5 m high, able to establish and grow rapidly on intertidal mudflats. The Grey Mangrove is a quick growing and hardy plant and quickly colonises new sites. Rapid colonisation of the shifting substrate is assisted by a seed which begins to germinate still attached to the parent tree, and can grow quickly when lodged in a suitable site by water movement. Survival on mudflats with daily inundation is enabled by pneumatophores, vertical projections from the roots which stand above the mud to allow oxygen intake in the normally anaerobic conditions. Increases in siltation in parts of Port Jackson and particularly in the Lane Cove River have resulted in some increases in the size of mangrove stands, generally at the expense of saltmarsh areas[26].

A second, less common, mangrove is River Mangrove, *Aegiceras corniculatum*, a less vigorous species. It grows to a shrub about 2 m high and does not appear to favour fully saline sites, being generally confined to the brackish end of the estuaries, often further upstream than Grey Mangrove. Where the two species overlap the River Mangrove is found as a band on the landward side.

Where the shoreline is rocky and the estuary is deep, as along the northern side of the Parramatta River and in Middle Harbour, mangroves, if they can establish a hold, are restricted to a narrow line immediately along the shore. Where there are more extensive mudflats and alluvial depositions, such as along the more gently sloping shoreline of the southern side of the Parramatta River or Botany Bay, zones of saltmarsh and *Juncus* meadow may be found between the mangroves and the shoreline.

Saltmarsh is a low-growing herbland characterised by mats of the succulent-stemmed Samphire or Glasswort, *Sarcocornia quinqueflora*, and the taller growing Seablite, *Suaeda australis*. Both species are in the family Chenopodiaceae, many members of which are able to cope with high levels of salt in the environment and with regular periodic flooding. Saltmarsh areas are flooded less frequently than their mangrove margins but may have the highest soil salinity, resulting from the high evaporation rate following very shallow flooding by the sea. Sites around the harbour and at Homebush Bay were used for salt collection in the early nineteenth century.

Occasionally flooded and rarely flooded sites have grassland of *Sporobolus* and *Zoysia*, or *Juncus* meadow with *Juncus kraussii*, while beyond normal tidal influence is Swamp Oak *Casuarina glauca* woodland with occasional shrubby paperbarks, *Melaleuca ericifolia*, with a ground layer of *Baumea juncea* and *Juncus kraussii*. Such areas still occur on the Georges River near East Hills.

Estuarine Wetlands have generally survived much better than their freshwater counterparts, simply because they were not suitable for anything but the roughest agriculture. Most destruction has been by landfill tipping and the construction of playing fields.

4 The European Impact

Two centuries of change

Some of the actual plant specimens collected by Joseph Banks and Daniel Solander at Botany Bay are now held by the National Herbarium of New South Wales, in the Royal Botanic Gardens. Here a specimen of *Banksia integrifolia* still retains the label written by Banks' botanist-librarian, Jonas Dryander. (J. Plaza, RBG, 1990)

Early botanical work in Sydney

For the Aboriginal people of the Great South Land, and its plants and animals, the *Endeavour's* visit to Botany Bay in 1770 was to have no repercussions for 18 years. However, for the two botanists accompanying Captain Cook — Dr Daniel Solander and the young Joseph Banks — there were exciting scientific discoveries to be made, and Banks spent much of his time 'in the woods botanizing as usual'[1]. Although we are not sure of the number of plant specimens collected, as these were subsequently distributed to various institutions, the collection included over 80 new species and gave rise to 94 coloured sketches.

Banks' enthusiasm for Australia never waned, and it was chiefly on his recommendations that Botany Bay was chosen as the site for a proposed self-supporting penal settlement. A man of great ability as well as a close personal friend of most of the important people in the British government, Banks was consulted on matters relating to the administration of the new colony for 30 years. It is strange, therefore, that no qualified botanist, horticulturist or agriculturist was included with the First Fleet. Governor Phillip was to report to Lord Sydney in May 1788:

> and here, my Lord, I must beg leave to observe, with regret, that being myself without the smallest knowledge of botany, I am without one botanist, or even an intelligent gardener, in the colony; it is not therefore in my power to give more than a very superficial account of the produce of this country, which has such variety of plants that I cannot, with all my ignorance, help being convinced that it merits the attention of the naturalist and the botanist.[2]

There seems to be no reason for the omission of a

botanist in the generally well-organised First Fleet complement.

The place of the naturalist appears to have been filled by some of the officers of the fleet. The surgeon-general, John White, was interested in the flora and fauna and his *Journal of a voyage to New South Wales*, published in London in 1790[3], included descriptions of plants and animals and 65 copper-plate engravings of birds, animals and botanical specimens. His botanical collections, sent back to England, provided the type specimens for the botanist Sir James Edward Smith to describe the tree species, *Eucalyptus capitellata, Eucalyptus haemastoma, Eucalyptus pilularis, Eucalyptus piperita, Eucalyptus resinifera, Eucalyptus robusta, Eucalyptus saligna* and *Eucalyptus tereticornis,* all found close to Sydney Cove.

One of White's medical assistants, Denis Considen, sent herbarium specimens and grass tree gum to Sir Joseph Banks and would appear to be the founder of the eucalyptus oil industry. Writing to Banks in November 1788:

> We have a large peppermint tree which is equal to, if not superior, to our English peppermint. I have sent you a specimen of it. If there is any merit in applying these and many other simples to the benefit of the poor wretches here, I certainly claim it, being the first who discovered and recommended them.[4]

The first properly trained botanical collector, George Caley, arrived in Sydney in 1800. Banks paid him a salary (15 shillings a week) and the government provided a house and rations at Parramatta. Here a botanical garden was established and Caley named as Superintendent. By 1803 he had made a number of collecting excursions around Sydney, to the Nepean, north of Camden; south of Camden; to The Oaks; and to Bents Basin and Picton (Thirlmere) Lakes. In 1804 he almost succeeded in crossing the Blue Mountains, getting as far as Mt Banks before being forced to return. Another trip was 'to the Sea' — from Pennant Hills to Cabbage Tree Lagoon (now Narrabeen Lake).

In 1802 another of Banks' botanists, Robert Brown, arrived with Matthew Flinders aboard the *Investigator* to explore the coasts of Australia. He spent some time in the Sydney area, working with Caley, but left no descriptions of the vegetation. As the leading botanist of his day he prepared most of the original descriptions of a vast number of plant species, and his *Prodromus Florae Novae Hollandiae* (1810) was the first systematic account of the Australian flora. J. D. Hooker, later Director of Kew Gardens, wrote that he 'united a thorough knowledge of the botany of his day with excellent powers of observation, consummate sagacity, an unerring memory, and indefatigable zeal and industry as a collector and investigator'[5].

The diligent collecting and time-consuming pressing of so many plant specimens by the botanists did not go unnoticed:

> Though thousands of thy vegetative works
> Have, by the hand of Science (as 'tis called)
> Been murder'd and dissected, press'd and dried,
> Till all their blood and beauty are extinct;
> And nam'd in barb'rous Latin, men's surnames,
> With terminations of the Roman tongue;
> . . . Poets are few,
> And botanists are many, and good cheap

satirised Barron Field[6] in 1825, in *First Fruits of Australian Poetry*.

Allan Cunningham, Caley's successor, arrived in Sydney in 1816. He was attached to John Oxley's 1817 expedition as botanist, and then accompanied Phillip Parker King's hydrographic survey of the north and north-west coasts of Australia. In 1822 he resumed botanical collecting in New South Wales, visiting the Illawarra, the Blue Mountains and the Bathurst–Mudgee districts, to be followed by explorations to the Liverpool Range south of Tamworth, and the Darling Downs in south-east Queensland. His field notes and specimens are at Kew Gardens and the British Museum of Natural History. Appointed Colonial Botanist and Superintendent of the Botanic Gardens in Sydney in 1837, he resigned within a year because supervisory duties left him little time for research. The ledger stone from his original grave in the old Devonshire Street cemetery is preserved in the wall of the National Herbarium of New South Wales' Robert Brown Building in the Royal Botanic Gardens.

Following Cunningham's death in 1839, active study of the Australian flora moved from Sydney back to Kew in London, where George Bentham was preparing his monumental *Flora Australiensis*, of seven volumes appearing between 1863 and 1878. He was assisted by the Director of the Melbourne Botanic Gardens, Ferdinand von Mueller, the outstanding colonial botanist of the second half of the nineteenth century and the focal point of botany in Australia at the time. Sydney did not re-emerge as a centre for botanical research until after the establishment of the National Herbarium of New South Wales at the Sydney Botanic Gardens in the 1880s. Joseph Henry Maiden, Director from 1896 to 1924, increased the collection of Australian plant specimens in Sydney, and indeed most of the older specimens available now date from his period.

Since that time the collection of the National Herbarium has been continually enlarged, making it an invaluable resource for botanical research in Australia. However, when trying to establish the original distribution patterns for some species it is disappointing to find that most of the early specimens are still in European herbaria.

The impact of agriculture

In 1779 Joseph Banks gave evidence before a House of Commons Select Committee investigating the possibility of setting up overseas colonies. He described the Australian east coast:

The Proportion of rich Soil was small in Comparison to the barren, but sufficient to support a very large Number of People; . . . the Grass was long and luxuriant, and there were some eatable Vegetables, particularly a Sort of Wild Spinage; the Country was well supplied with Water; there was an abundance of Timber and Fuel.[7]

Nine years later Governor Phillip's impressions of Botany Bay were very different. The grasslands turned out to be swampy heaths, the timber proved to be of poor quality, and there was not enough fresh water for a substantial settlement (Banks saw the country in late April whereas Phillip arrived in midsummer).

A week later Phillip moved the fleet up to Port Jackson. The harbour foreshores enchanted the new arrivals. To Surgeon Worgan, Port Jackson suggested:

to the Imagination Ideas of luxuriant Vegetation and rural Scenery, consisting of gentle risings & Depressions, beautifully clothed with variety of Verdures of Evergreens, forming dense Thickets, & lofty Trees appearing above these again, and now & then a pleasant checquered Glade opens to your View. — Here, a romantic rocky, craggy Precipice over which, a little purling stream makes a Cascade. There, a soft vivid-green, shady Lawn attracts your Eye.[8]

Once the soldiers and convicts had disembarked, however, there was the serious business of establishing a colony in this new continent: 'the principal Business has been the clearing of Land, cutting, Grubbing and burning down Trees, sawing up Timber & Plank for Building, making Bricks, hewing Stone, Erecting temporary store-houses, a Building for an Hospital, another for an Observatory, Enclosing Farms & Gardens,' wrote Surgeon Worgan.

'Except for the size of the trees, the difficulties of clearing the land are not numerous, underwood being rarely found, though the country is not absolutely without it,' reported Watkin Tench[9]; he was evidently not involved in the actual tree removal. John White writes:

It will scarcely be credited when I declare that I have known twelve men employed for five days in grubbing up one tree; and, when this has been effected, the timber (as already observed) has been only fit for firewood; so that in consequence of the great labour in clearing of the ground and the weak state of the people to which may be added the scarcity of tools, most of those we had being either worn out by the hardness of the timber or lost in the woods among the grass through the Carelessness of the convicts, the prospect before us is not of the most pleasing kind.[3]

The cove to the east of the settlement was chosen for a farm as it seemed a convenient, relatively level site with a stream of fresh water and where 'the trees stand at a considerable distance from each other'[10]. Woccanmagully of the Aborigines had become Farm Cove.

Unfortunately, the sandy soils were soon found to be very infertile. 'I do not scruple to pronounce that in the whole world there is not a worse country than what we have yet seen of this,' wrote Lieutenant-Governor Ross[11], not the most enthusiastic of the officers. Eight months after the landing even Phillip's optimistic reports on the farm's progress were starting to wane:

It is now found that very little of the English wheat had vegetated, and a very considerable quantity of the barley and many seeds had rotted in the ground...All the barley and wheat likewise which had been put on board the *Supply* at the Cape were destroyed by the weevil. The ground was, therefore, necessarily sown a second time with the seed which I had saved for the next year, in case the crops in the ground met with any accident.[12]

Only a miserable crop of about a bushel was harvested from the first six acres at Farm Cove, an extremely poor result considering that about 12 bushels were sown.

The settlement was saved by the discovery of better farming land further west, at Rose Hill near Parramatta on the edge of the Cumberland Plain. Here by December 1789, 200 bushels of wheat, about 35 bushels of barley, and a small quantity of oats and

Broad-leaved Apple, *Angophora subvelutina*, indicated rich floodplain soils. These were rapidly cleared for agriculture by the early settlers. A remnant tree, probably somewhere near Penrith or Camden, is featured in this glass slide photograph used to illustrate botany lectures at the turn of the century.

Another turn of the century photograph, but of a young Grey Box tree, *Eucalyptus moluccana*, characteristic of woodland on the clay soils of the Cumberland Plain. Today's vegetation is mostly of similar-sized, or smaller trees that have regrown in sites where pressures from grazing have been reduced.

maize had been produced, compared with 25 bushels of barley from Farm Cove that season. To demonstrate the possibility of farming self-sufficiency, a grant of 30 acres [12 ha] of nearby land was given to the ex-convict James Ruse, who had had practical farming experience in England. On 'Experiment Farm' he put into practice techniques that led to successful farming in the colony.

Agriculture expanded, particularly with settlement of the fertile floodplain of the Hawkesbury River near Windsor in the 1790s. 'It is this river, whether we call it Hawkesbury or Nepean, that is the Nile of Botany Bay: for the land on its banks owes its fertility to the floods which come down from the Blue Mountains, and which have been known to swell the waters nearly a hundred feet [30 m] above their usual level,' wrote Barron Field[13] in 1822. The floods on the Hawkesbury remain a feature of that environment, though the water-storage dams upstream have altered the character of the flooding.

Clearing of the alluvial flats proceeded rapidly. Indeed problems of erosion and flooding were soon apparent. In 1803 Governor King wrote to Lord Hobart:

> From the improvident method taken by the first settlers on the sides of the Hawkesbury and creeks in cutting down timber and cultivating the banks, many acres of ground have been removed, lands inundated, houses, stacks of wheat, and stock washed away by former floods, which might have been prevented in some measure if the trees and other native plants had been suffered to remain, and instead of cutting any down to have planted others to bind the soil of the banks closer, and render them less liable to be carried away by every inconsiderable flood.[14]

King issued orders prohibiting the destruction of 'any tree or shrub growing within two rods [10 m] of the edge of the bank', and 'earnestly recommended to those who already hold farms by grant situated on the side of any river or creek liable to floods, and which have been cleared of timber, to replant the banks with such binding plants and trees as they can procure'. Doubtless little was done, for 20 years later Barron Field was writing:

Another good reason against granting away this land, and suffering it to be cleared, is, that the floods wash the fallen timber into the channel of the river, and obstruct the navigation. The removal of the trees from its banks has not only contributed to choke the river by their falling in, but has occasioned derelictions on one side and alluvions on the other.[13]

At the same time grazing runs were being taken up on the woodlands of the Cumberland Plain. In June 1788, seven head of cattle strayed from the herdsman at Sydney and made their way to the south-west. Here on 'The Cowpastures' they were discovered by John Hunter in 1795. Because of the importance of the cattle to the early colony, successive governors proclaimed the area 'off limits' to settlement. The ambitious John Macarthur, however, keen to develop his growing flocks of sheep, argued that this was the only suitable land for sheep in the County of Cumberland, and managed (with the patronage of Lord Camden) to obtain a grant of 2,000 hectares over much of The Cowpastures. 'Camden Park' came to epitomise the aspirations of the nineteenth century colonial gentry.

Elsewhere on the Cumberland Plain, rural activity was proceeding rapidly. By the 1820s James Atkinson could write:

In the county of Cumberland, one immense tract of forest land extends, with little interruption, from below Windsor on the Hawkesbury, to Appin, a distance of 50 miles [80 km]; large portions of this are cleared and under cultivation, and of the remainder that is still in a state of nature, a great part is capable of much improvement. The whole of this tract, and indeed all the forest in this county, was thick forest land, covered with very heavy timber, chiefly iron and stringy bark, box, blue and other gums, and mahogany.

He advised the settlers that 'the best forest lands are invariably thinnest of trees; and in general it will be found the best lands are the least encumbered with timber'. The importance of the geology in determining the soil type and subsequent vegetation was also recognised:

The quality of forest land, and indeed of most others will be found to be governed by the nature of the rocks and stones that form the basis of the soils; thus, in this tract of forest, in the county of Cumberland, the rocks are either common or calcareous sandstone, ironstone, and in some few places, whinstone [probably basaltic]; these form soils of various degrees of goodness, the whinstone generally the best.[15]

Wheat was being grown near Campbelltown, orchards and vineyards were being established, and clearing for grazing was taking place everywhere. The native grasses, evolved in the absence of hard-hooved grazers, had not experienced such grazing pressures as that provided by the introduced sheep and cattle, and by the 1820s overgrazing was becoming a serious problem. It was only by moving stock across the Blue Mountains to the 'unlimited' western pastures that more sustainable stocking rates were achieved.

Louisa Meredith complained about the rate of clearing in 1839:

The system of 'clearing' here, by the total destuction of every native tree and shrub, gives a most bare, raw, and ugly appearance to a new place. In England we plant groves and woods, and think our country residences unfinished and incomplete without them; but here the exact contrary is the case, and unless a settler can see an expanse of bare, naked, unvaried, shadeless, dry, dusty land spread all round him, he fancies his dwelling 'wild and uncivilized'.[16]

By the late nineteenth century western Sydney's landscape had changed irrevocably. Apart from the obvious destruction of the forests and woodlands, we know little of the detailed losses of plant and animal species. No records were kept. The relative abundance of native species changed, with a few favoured by the new conditions becoming more common but most becoming rarer.

Changing attitudes

Since Banks and Solander's visit to Botany Bay in 1770, botanical visitors have been fascinated by the richness and diversity of Sydney's plants. But despite the initial optimism of the First Fleet writers, the surrounding landscape and vegetation was often seen as alien and monotonous. For example, Charles Darwin, after travelling from Parramatta to Penrith, took the opportunity to write:

The extreme uniformity of the vegetation is the most remarkable feature in the landscape of the greater part of New South Wales. Everywhere we have an open woodland; the ground being partially covered with a very thin pasture. The trees nearly all belong to one family [Myrtaceae]; and mostly have the surface of their leaves placed in a vertical, instead of as in Europe, a nearly horizontal position: the foliage is scanty, and of a peculiar, pale green tint, without any gloss. Hence the woods appear light and shadowless... The greater number of the trees, with the exception of some of the blue gums, do not attain a large size; but they grow tall

Fig. 4 This early scene of Brickfield Hill (today George Street near Liverpool Street), shows the almost unmistakable pink-blotched trunk and twisted branches of *Angophora costata*. In the distance the forest is being pushed back though the Europeans have only been here for nine years. (Dayes, Edward, 1763–1804, *Brickfield Hill and village on the High Road to Parramatta* (ca.1797), watercolour, National Library of Australia)

Fig. 5 Sydney Cove in 1788, as seen by Lieutenant William Bradley, shows a rocky foreshore with forest beyond. (Mitchell Library, State Library of New South Wales)

Fig. 6 Turpentine-Ironbark Forest in the Wallumetta Forest at Twin Road, Ryde.

and tolerably straight, and stand well apart. The bark of some falls annually, or hangs dead in long shreds, which swing about with the wind; and hence the woods appear desolate and untidy. Nowhere is there an appearance of verdure, but rather that of arid sterility.[17]

Perhaps only Caley could see positive qualities in the landscape; his place names — Gaping Gill, Dismal Dingle, The Cataract of Carrung-Gurrung, Moowat'tin and Dovedale — demonstrate strong feelings; his acceptance of Aboriginal names indicates a closer relationship with the country than was customary. Few Aboriginal plant names have survived, despite the interest in the flora — Waratah, Bangalay, Burrawang, Geebung, Kurrajong and Gymea are the only ones still in use; none was ever incorporated into scientific nomenclature.

During the second half of the nineteenth century, as much of the bush was disappearing before the axe and saw, an increasing awareness of the colony's natural attributes developed, undoubtedly encouraged by the Great Exhibitions, a feature of the Victorian period — particularly the Sydney Exhibition of 1879 and Melbourne Exhibition of 1882. Guide books such as Gibbs, Shallard and Co's *Illustrated Guide to Sydney* (1882) could now assert under 'Hints for Tourists':

Tis true we have no mouldering monasteries, time-honoured cathedrals, or moss and ivy-clad dilapidated castles, rendered venerable and interesting by historical association. We do not see the cowslip in our meadows, and our interminable forest shades know but little of the sweet songsters of the grove; but we have a fauna extensive and unique, a brilliant sunshine, a clear blue sky, gorgeous flowering shrubs and plants; vast wooded ranges, deep solemn glens and mountain gorges, which, together with the indescribable air of intensified solitude, which particularly distinguishes our Australian scenery, are amply sufficient for poetical inspiration.[18]

The increasing wealth of the colony led to a demand for decorative objects. Sydney's bush plants now provided a rich source of ideas for design motifs in decoration and ornament. Waratahs, Sydney Wattle, Flannel Flowers, Christmas Bells, Native Rose and Native Heath appeared as decoration on furniture, silver, porcelain, leather and particularly in stained-glass windows. Plasterwork and pressed-metal ceilings carried ornate wildflower decorations. The 1888 Centenary and Federation in 1901 provided high points for 'Australiana' decoration. The creation of the Gumnut Children and the Big Bad Banksia Men by May Gibbs continued this tradition into the twentieth century[19].

'On the heights we fill our hands with wild roses, mountain moss, and waratahs'[18]. Decorating with wildflowers was popular as Sydneysiders grew more interested in their natural surroundings. (Macleay Museum)

Forest on the shale soils north and south of Sydney Harbour provided timber for constructing buildings, bridges and wharves. Smaller trees were cut for firewood. Where uncleared, such forest areas were grazed, and ground and shrub species disappeared. The coming of the railway made such areas accessible for suburban living, and housing subdivisions like this one at Wollstonecraft followed. (Mitchell Library, State Library of New South Wales)

'the overflow of bricks and mortar'

As early as 1856 the naturalist, the Rev. William Woolls of Parramatta, was concerned about the loss of native species. Woolls pointed out that:

> owing to changes which are taking place amongst us the 'native botany' will soon become a matter of history. Perhaps some of your readers may smile at this, but nevertheless it is a fact, for owing to the destruction of native plants on the one side of Parramatta by the waterworks, and on the other side by the railway, I fear that several of the Myrtaceae [Eucalypt family] and Leguminosae [Pea family] have already disappeared.[20]

In the second half of the nineteenth century, Sydney started to acquire its suburban character. The wealth acquired in the gold rushes and from the developing agricultural industries, particularly wool, led to an expanding population and concurrent housing boom. As the pace of agricultural expansion eased, Sydney's surviving bush began to face increased pressures from the spread of the suburbs. Such development destroyed every component of the bushland except perhaps the odd tree or remnant in an unsuitable building site. The development of mechanised transport — trains in the 1850s, trams and ferries in the 1890s, and buses and cars after World War I — promoted the continual spread of the suburbs.

The construction of the railway from Sydney to Parramatta in 1855 began the changes on the rural Cumberland Plain. The railway gave impetus to industry, and later to speculative land subdivisions. By the 1880s the western railway lines had attracted a number of industries, including sawmills which cut the eucalypts and she-oaks for railway sleepers and firewood, and tanneries which used the bark from local wattles — Green Wattle, *Acacia decurrens,* and *Acacia parramattensis* — to make tanning solutions for leather. The area between Prospect and Penrith became a supplier of clay for bricks, sand and gravel for cement, and from the 1880s blue metal for roads. These resources were being consumed by the growing suburbs of the city.

Meanwhile, in the suburbs, between Newtown and Petersham, 'houses, villas, gardens, and slowly developing streets are successively presented; where (not long since) there was nothing but open country, or shady "bush",' reported the 1879 *Railway Guide*:

> The first clump of forest trees, yet undisturbed, on the old Annandale Estate, next shows itself on the right, and is, of course, the object of much curious speculation to European visitors, unaccustomed to the rather stiff and formal eucalyptus. On approaching Petersham Station a fine view over the country unfolds itself to the right — the celebrated 'Blue Mountains' becoming visible far away to the westward. Petersham Station is now the centre of a thickly populated suburban district, and on the slopes around it are many really delightful villas and gardens — Usual time of *trajet* from Sydney to Petersham, about 12 minutes.[21]

In the 1880s subdivision reached a peak, and many estates were cut up — Annandale, Arncliffe, Paddington, Newtown, Eveleigh, Rose Bay, Woollahra, Canterbury, Marrickville, Ashfield, Rockdale, Randwick, Stanmore, Leichhardt and Forest Lodge. In 1883 Sydney's population reached 225,000. So rapid was the building boom that

> the overflow of bricks and mortar has spread like a lava flood over the adjacent slopes, heights and valleys, until the houses now lie, pile on pile, tier on tier and succeed each other row after row, street after street, far into the surrounding country and the eruption is still in active play and everywhere the work of building and expansion proceeds at a rapid pace.[22]

Within ten years the population had nearly doubled.

The nineteenth century suburbs followed the transport corridors, the railway from Redfern to Burwood; the tramlines to Randwick, Paddington, Annandale, Waverley and Woollahra; and the ferry routes to Balmain, Rose Bay, North Sydney and Mosman. Most of the developing suburbs were on the level, already-cleared shale country. Where rugged sandstone hillsides were too conveniently located to be overlooked, such as at Balmain, Glebe or Paddington, the 'terrace' style of architecture was ideal for stepping down steep slopes.

After Federation, extension of the public transport system, particularly the tram network, made new areas accessible. Single-storey houses on larger blocks of land — the Federation house in the garden suburb — became fashionable. Federation suburbs were mainly on the flatter shale areas — Haberfield, Croydon, Burwood and Chatswood, for example — but also at Rockdale, Randwick and Mosman.

Between the two World Wars, the increasing availablity of motor cars allowed the development of suburbs away from the main public transport network. Most of this expansion was into land previously cleared for agriculture, west of Sydney, from Canterbury to Bankstown, Concord to

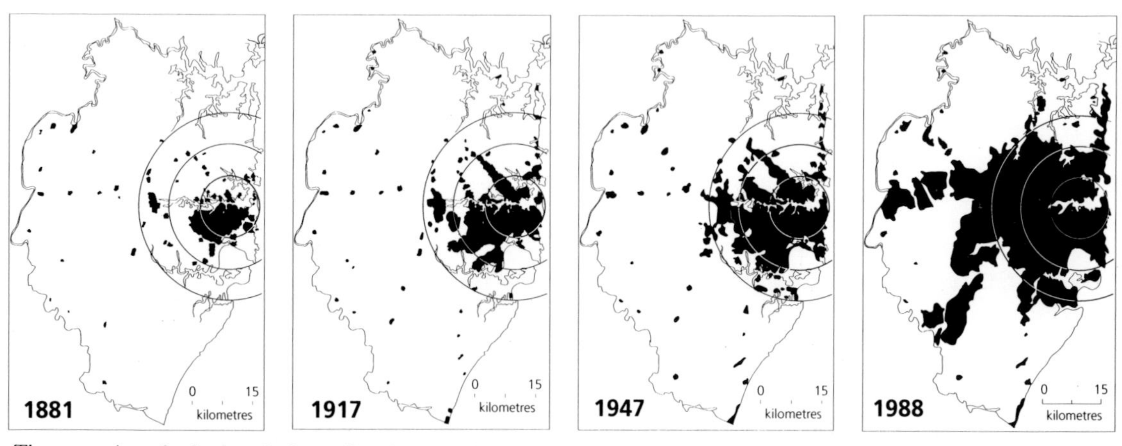

The expansion of suburban Sydney. (Cumberland County Council, 1948 and Dept of Environment and Planning, 1988)

Suburban housing begins to change the face of former farm and forest land in Eastwood early this century. (Photo George Hawkins, 1912. Ryde Municipal Council Library Service, Local Studies Collection)

Parramatta, the North Shore and the Warringah area. 'Willoughby spreads in the unrelieved ugliness of closely packed houses towards Castlecrag,' wrote I. Anson in September 1927[22]. The opening of the East Hills railway line in 1931 promoted major suburban development in the south-west.

However, it was not until after World War II that housing spread from the main ridges onto the steeper, uncleared sandstone hillsides. Expansion on the North Shore — at Lane Cove, Pymble, Ryde, around Middle Harbour, Frenchs Forest and along the northern beaches — destroyed much bushland. In the south, bushland along the Georges River met a similar fate. In the past decade extensive bushland areas have been destroyed at Menai, Campbelltown and Berowra, for example. Even the bushland that remains in parks and reserves near these housing areas is threatened as nutrient-enriched run-off and changed fire regimes allow exotic weed species to invade.

Cyclic and irreversible change — the obvious and the imperceptible

Bushland is not a static arrangement of plants, as has perhaps been suggested by our descriptions. While trees, shrubs, grasses, herbs or vines of many different species may grow together, there are differences in their requirements for light, nutrients and water, their ways of dispersing seeds and their growth rates and life spans. For example, while some trees may take 50 to 100 years to mature, some shrubs may be dying of old age after 25 years or herbs after several years.

Individual plants grow, flower and shed seed. Plants age and die, to be replaced by others, either from seed or by vegetative growth. Changes in a plant community may be cyclic or directional. Broadly speaking, natural cycles in the plant community may be maintained by recurring events such as fire or flood. After these, species already in the habitat, perhaps as soil-stored seed, are recruited and the vegetation gradually regains its former structure. Depending on the sequence of events and weather conditions, slightly different combinations of species may be favoured, but the plant community is essentially self-maintaining. Directional change may be induced by habitat disturbance which alters soil and water properties, and allows recruitment of species not native to the habitat, in particular, exotic weeds. Such changes generally lead to bushland deterioration.

Let us look at some of the factors important in change in the plant community.

Soil

The soil provides nutrients and water, as well as providing physical support for the plant. Soils differ in the conditions they provide — in quantities of nutrients and physical properties such as drainage and depth. Different plant communities have developed on different soil types. In the Sydney area, the two main types of soil are those on the Hawkesbury Sandstone — sandy soils, low in nutrients and water-holding capacity — and those of the Wianamatta Shale — heavy clay soils of moderate fertility and higher water-holding capacity. Quite different plant communities grow on each.

Studies have shown that the type of plant community present is dependent upon the nutrient level of the soil and in particular on the level of phosphorus. The typical vegetation of the low-phosphorus sandstone soils has many sclerophyllous (hard-leaved) shrubby species, whilst shale soils have more mesophyllous (soft-leaved) shrub species, and a higher proportion of grasses and herbs. On soils with the highest nutrient levels, the alluvium with River-flat Forests, a considerable number of rainforest species occurred.

Where nutrients are added to low-nutrient soils, changes are induced in the plant communities[23]. Phosphorus concentrations in urban stormwater are

Aerial photos in 1951 (left) and 1970 (right) show the impact of the suburban housing boom that followed World War II. Here at Killarney Heights ridgetop bushland bears the brunt. (Dept of Lands)

about 50–100 times greater than those that occur in natural streams in the Sydney region[24]. In northern Sydney, soil phorphorus levels have been found to be doubled at suburban boundaries, tripled in creek lines, and increased eight times below stormwater outlets[25]. Contributors are garden fertilisers, dumped refuse, sewer discharges, and, significantly, well over half comes from pet excrement. Pets may add from 3 to 10 kg per hectare per year of phosphorus to suburban areas[26]. Whilst some higher nutrient native plants might be able to take advantage of the changed conditions, for example, *Pittosporum undulatum*, the main benefactors are the exotic weed species. Weed propagules are dispersed into bushland in stormwater, dumped garden refuse, by wind or fruit-eating birds, and soon take advantage of any nutrient-enriched conditions. Increased soil phosphorus content has been shown to increase the weed component, and weed growth is particularly conspicuous along creeks and drainage lines below disturbed areas. Many creeks in and around Sydney have become overgrown tangles where Privet, Blackberry, Honeysuckle, Balloon Vine, and Morning Glory have replaced the native creekside vegetation of shrubs and small trees.

The addition of nutrients to the soils of native plant communities begins changes that are irreversible. Protection of bushland from these changes depends on protection of catchments from disturbance. Where suburban development has taken place, the addition of the nutrients is an ongoing process. Nutrient input may be reduced by careful design or redirection of stormwater drains[24 27], though in practice this may be very difficult, and once the nutrients have entered the system they are virtually impossible to remove.

Fire

The ancestors of Sydney's bushland plants evolved mechanisms that enabled them to survive fire over many millions of years. With Aboriginal occupation, fire frequency appears to have increased and patterns resulting from the inevitable wildfires, superimposed on the localised burns of the Aborigines, increased the diversity of habitats inherent in the sandstone and shale landscapes.

With the arrival of the Europeans, fire patterns were again altered. Fire was either excluded completely for long periods, or used more frequently to clear fuel. Because of their size and location, today's suburban bushland reserves tend to be subjected to very frequent fires intended to keep fuel loads low, or to total fire exclusion. These treatments alter the diversity of species in different ways.

Modern housing continues to destroy bushland. Careful planning by local government and service authorities can minimise the adverse effects of continuing suburban expansion on bushland remnants. (1988)

Bushland plants survive fires in a number of ways. Some plants — 'obligate seeders' — are killed, but their seeds are protected so that the next generation can begin after a fire. Seeds may be protected in woody cones, as in *Banksia* and *Hakea* species, or with hard seed coats, as in many wattles and peas, that protect seeds lying dormant in the soil. Fire opens the cones and cracks the hard seed coats to allow germination of seedlings[28]. Other plants — 'resprouters' — survive fire as adults, regrowing from protected buds or underground parts. 'Epicormic' shoots along eucalypt trunks and branches, and resprouting shrubs, grasses and herbs, soon turn blackened bushland green.

Frequent fires may kill obligate seeders before they mature and set seed, causing local extinctions and allowing grasses, ferns and other resprouting plants to take their place[29 30 31]. Fire intensities also influence plant responses. Low intensity fires may not be hot enough to stimulate germination of seeds lying dormant in the soil. Changes induced by cool season fires may therefore differ from those following fires during hot seasons. Today episodic hot crown fires, once a significant event in the maintenance of bushland, rarely occur because of small size of reserves and proximity of housing. In their absence, certain species, once a minor component of the bushland, will be favoured at the expense of others which were previously more common. Areas left unburnt for 30 years or more develop dense shrub layers, favouring more mesic (moisture-loving) species, such as *Pittosporum undulatum*. Subsequent growth of these species may inhibit regeneration of light-requiring plants, changing species composition of the plant community.

Flood

Floods affect a much smaller amount of bushland than fire. Major floods inundate the floodplains of the major rivers, the Nepean–Hawkesbury and the Georges, though the patterns have been altered by the construction of dams upstream, which have tended to reduce small flood occurrences, whilst having little effect on larger floods.

Native vegetation in flood-prone areas is adapted to withstand periodic flooding, and indeed may require it. Low-lying estuarine and saltmarsh areas with Swamp Oaks, *Casuarina glauca*, for instance, may be flooded by a combination of high tides and river floodwaters.

The main effects of flooding today are changes in the quality of the water and the amount of silt and weed seed material transported. Most floodwaters now contain quantities of nutrients and silt, as well as seeds and propagules of weeds and native species. The nutrients and silt enable the weed species to grow more vigorously and outcompete native species. In high-nutrient environments, as along the banks of

The Nepean River flooding *Casuarina cunninghamiana* and *Eucalyptus* trees at Menangle, south of Camden in 1973. Flooding disperses propagules of both native and exotic species, though the exotic species grow most vigorously on the recently deposited stilt.

the Nepean River, species of Privet are major invaders, but there are a considerable number of other weeds of horticultural origin. A great variety of plants was grown at 'Camden Park' in the nineteenth century, and their seeds have evidently been spread along the Nepean during floods. In nutrient-enriched sandstone gullies draining urban housing, such as Berowra Creek, Privet, Wandering Jew and Crofton Weed are establishing rapidly.

Drought

There is little information on the effects of prolonged drought on Sydney's vegetation. In some instances it may initiate extra-vigorous flowering, or result in the death of occasional plants, such as tree ferns in moist gullies. Extensive plant death is rarely observed. The increased run-off associated with urban areas means that urban bushland may receive more water than in the past, reducing the effect of drought.

Observations in the central west of New South Wales during the 1982/83 drought, when much of the area had its driest period since rainfall records were started, indicate that major drought can induce many responses in the vegetation similar to those caused by fire. Eucalypt canopy can be shed, some trees may die, while others produce epicormic regrowth following drought-breaking rain; fruits on Proteaceae normally retained on the bush may be opened, releasing seed; heat in the top layers of the soil may induce the breaking of the hard testa surrounding leguminous seeds, allowing maximum germination when rains come; finally, the opening of the canopy and death of many mature plants can enable a prolific germination and new growth of many species if adequate rain follows the conclusion of the drought[32].

More recent concerns that weather patterns may be changing as a result of the 'greenhouse' effect have implicatons for changes in natural vegetation. Past climatic changes have occurred in plant communities by the sifting and movement of plant populations. Much of today's vegetation occurs as remnant islands surrounded by modified grazing and suburban lands. If species are lost as a result of climatic change, it is unlikely that other native species suited to the new conditions will be able to establish because of the lack of nearby propagule supply. Instead, exotic weed species will be well-placed to form the new vegetation. The only exceptions likely are if the new conditions are drier; native species may be better able to exploit dry conditions than exotic species.

Grazing

Grazing pressures on native vegetation have changed over the past two centuries. Kangaroos and other marsupials were probably concentrated on the grassy woodlands of the Cumberland Plain. Here the introduction of domestic stock has had a significant effect on remnant vegetation. *Themeda australis,* Kangaroo Grass, is selectively grazed and quickly disappears, though it seems to recolonise when grazing pressures are lowered. The shrubby *Bursaria spinosa* may be grazed at a young stage but is unpalatable as a larger bush, so that established thickets will remain in grazing land. Pasture improvement has resulted in the addition of species such as *Paspalum* and many herbs to the grassy understorey. Rabbits have also undoubtedly taken their toll of native vegetation, but we have little direct evidence of their specific impacts. Insects are major herbivores in natural systems. Their role in Sydney's bushland has been little studied, although the increasing occurrence of dieback in woodland trees in rural New South Wales has focused attention on their role in disturbed natural systems.

Whilst the loss of species is irreversible, removal of grazing pressures may enable rarer species to increase in abundance. At the Mount Annan Botanic Gardens at Campbelltown, previously grazed woodland areas are being monitored to assess changes in the absence of grazing.

5 The Vegetation of Your District

People today are becoming more aware of their environment, and the need to maintain its quality. Bushland areas are being recognised as contributing to the quality of our urban environment, and there is an increasing demand for information on various aspects of Sydney's native vegetation. Information sought is not confined to the scenic tracks in the major national parks, but includes the types of plants found in local council reserves or the problems of conservation and management of bushland on undeveloped lands and in remnants along roadsides. People are also interested in what originally grew in the now well-established suburbs. These demands are a response not only to the increasing environmental content of school and tertiary education courses, but to the growing evidence of the impact of 200 years of European occupation on our environment, and the obvious losses or changes in bushland around us — bushland that we have generally taken for granted.

Because many of the enquiries that we receive in our work at the Royal Botanic Gardens relate to specific localities, we thought it appropriate to look at the vegetation of each of the 40 local government areas within the County of Cumberland. We have not specifically described the vegetation of the major national parks, Ku-ring-gai, Royal and Marramarra; they deserve books of their own, though much of what we describe of plant communities and interactions between the suburb and the bush is relevant to them. By providing a brief description of the original vegetation, as far as historical records and the examination of bush remnants permit us to reconstruct, we hope to give an idea of the past interactions between the bush and the development of the suburb, and the future management or rehabilitation of areas where necessary.

Local government areas do not always follow natural boundaries, and duplication of some information may occur. We have tried to reduce this by referring the reader to the previous descriptions of the eight main plant communities. Nor can the status and composition of all areas of remaining bush be given in detail. This is beyond the scope of this work. For people interested in specific areas, the References will provide access to more detailed information.

For convenience we have grouped the local government areas into five major geographic regions, and within these they are generally arranged alphabetically: (1) The City, Inner West and Southwest; (2) Parramatta and the Cumberland Plain; (3) the Eastern Suburbs; 4) North of the Harbour; and (5) the Southern Suburbs. Maps showing the distribution of Sydney's vegetation types in 1788 and 1990, and local government boundaries, appear on the endpapers.

The City, Inner West and Southwest

This section begins with the vegetation of Sydney Cove, the site of the first European Settlement, and then covers the inner central local government areas from the Central Business District to Auburn and southwest to the Georges River. The area is a reasonably cohesive geographically. It is mostly underlain by gently undulating Wianamatta Shale. Hawkesbury Sandstone outcrops along the edges of the Parramatta and Georges Rivers and near the estuary of the Cooks River, and there are estuarine alluvial deposits associated with these waterways. A gradient of decreasing rainfall from east to west appears to have influenced the original vegetation patterns.

1 The Sydney City Area

It is almost impossible to imagine the quiet wooded valley of the Tank Stream as it flowed into Sydney Cove in January 1788. The scale of the Central Business District's massive office towers has almost completely obliterated the original landform.

After examining a number of coves for a suitable settlement site, Governor Phillip wrote: 'I fixed on the one that had the best spring of water, and in which the ships can anchor so close to the shore that at a very small expence quays may be made at which the largest ships may unload'[1]. The shoreline along Sydney Cove would have been of sandstone ledges and rocks, perhaps with small sandy beaches and a sandbank at the mouth of the Tank Stream. It is unlikely that mangroves were present, as Phillip would have mentioned them as indicative of mud flats and shallow water. The ease with which he suggests quays could be built indicates deep water close inshore. No mangroves are shown in a 1789 painting by William Bradley (Figure 5).

The 1788 shoreline of Sydney Cove, since filled-in for Circular Quay, extended back beyond Bulletin Place on the western side where the Tank Stream entered. The Tank Stream itself, the source of the fresh water necessary for the success of the settlement, arose in a broad swampy basin between Market and Park Streets, just south of Centrepoint Tower. Although no details of the original vegetation have survived, the soils and topography indicate the vegetation would have been an open woodland of spreading Scribbly Gums, *Eucalyptus racemosa,* with a shrubby understorey with *Leptospermum flavescens, Banksia oblongifolia* and *Callistemon citrinus*. From King Street to Bridge Street the Tank Stream appears to have followed a sandstone gully with thickets of mesic shrubs 6–8 m high. Here Lillypilly, *Acmena smithii,* Cheese Tree, *Glochidion ferdinandi*, Blueberry Ash, *Elaeocarpus reticulatus*, and the small fragrant tree *Synoum glandulosum* probably grew. There may have been Cabbage Palms, *Livistona australis*, that provided material for the first huts.

Larger trees along the stream would probably have been Forest Red Gums, *Eucalyptus tereticornis*, with a ground layer of ferns such as *Blechnum cartilagineum* and *Todea barbara*, and grasses and herbs. On the sandstone on the sides of the gully there would probably have been trees of Blackbutt, *Eucalyptus*

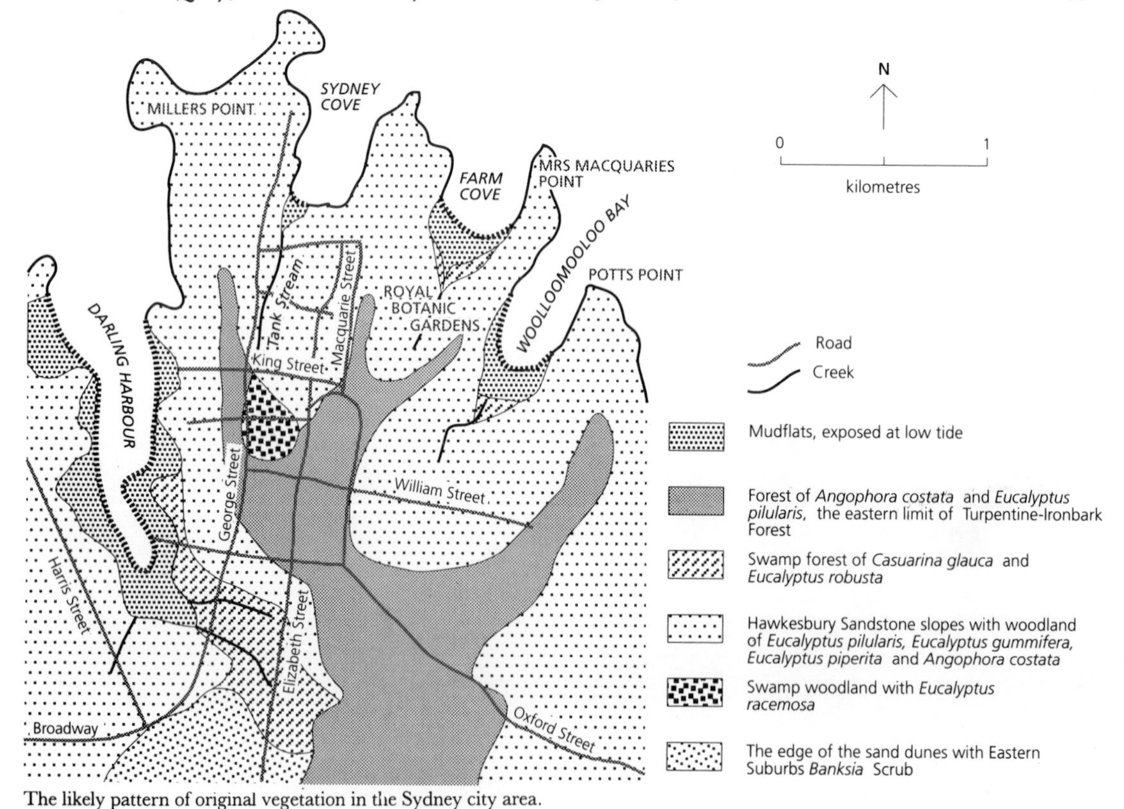

The likely pattern of original vegetation in the Sydney city area.

pilularis; Red Bloodwood, *Eucalyptus gummifera*; Sydney Peppermint, *Eucalyptus piperita*; and smooth-barked *Angophora costata.* Below Bridge Street the Tank Stream spilled out on to a small alluvial fan which probably had Swamp Mahogany, *Eucalyptus robusta*, on the higher parts and Swamp Oak, *Casuarina glauca*, nearer the salt water. Today Vaucluse Bay near Vaucluse House provides a similar appearance to the 1788 Sydney Cove, though on a smaller scale. It has a small sandy beach with Swamp Oak and Forest Red Gum on the flats and a small creek. On each side sandstone outcrops to form the shoreline, supporting a few old Port Jackson Figs, *Ficus rubiginosa*.

The hillsides away from the Tank Stream had shallow sandy soils developed from the Hawkesbury Sandstone. Governor Phillip wrote:

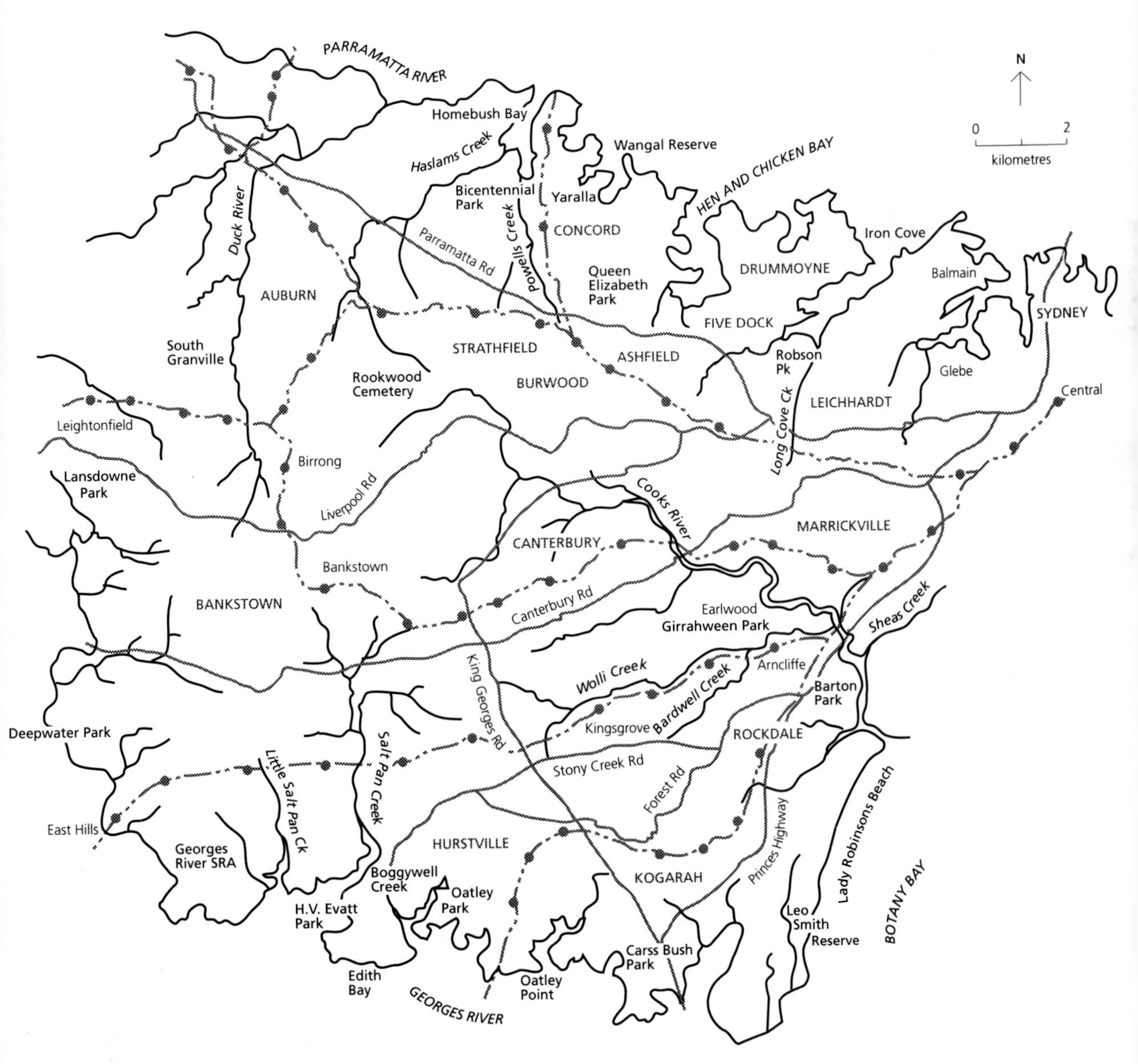

City and inner west sites mentioned in text.

The necks of land that form the different coves, and near the water for some distance, are in general so rocky that it is surprising such large trees should find sufficient nourishment, but the soil between the rocks is good, and the summits of the rocks, as well as the whole country round us, with few exceptions, are covered with trees, most of which are so large that the removing them off the ground after they are cut down is the greatest part of the labour.[1]

On the lower slopes we picture a woodland of Scribbly Gum, *Eucalyptus racemosa*, with trees up to 10 m high, with an open low shrubby understorey with *Acacia suaveolens, Banksia spinulosa,* and the tea-trees *Leptospermum attenuatum* and *Leptospermum flavescens*.

On the upper slopes and ridges would have been woodland of Blackbutt, *Eucalyptus pilularis,* and *Angophora costata* with a shrubby understorey including *Ceratopetalum gummiferum, Kunzea ambigua, Platysace lanceolata, Leptospermum attenuatum,* as well as *Dianella caerulea* and the grass *Themeda australis*. 'Between Sydney Cove and Botany Bay the first space is occupied by a wood, in some parts a mile and a half, in others three miles across; beyond that, is a kind of heath, poor, sandy and full of swamps,' reported Governor Phillip. The 'wood' would have been on the shale extending from Hyde Park to the southwest. At Brickfield Hill the deep clay was used for brick-making. Here was the eastern end of the Turpentine–Ironbark Forest, with *Eucalyptus pilularis, Eucalyptus resinifera, Eucalyptus paniculata* and *Angophora costata*. A 1796 painting of Brickfield Hill by Edward Dayes shows an almost unmistakeable *Angophora*, with its characteristically twisted branches and smooth, pinkish bark, together with the stump of a more questionable *Eucalyptus pilularis* in the distance (Figure 4). South of Central Railway were the sand dunes of the Eastern Suburbs with 'heath, poor, sandy and full of swamps'.

The clearing of trees to make way for the first huts marks the beginning of the displacement of the bushland by suburban Sydney. The city area was rapidly denuded; Dayes' painting shows few trees within the settlement, but on the horizon the featureless wall of bush is ready to be pushed back. Other early views show the town with virtually no native trees, or with perhaps a characteristically deteriorating tree for artistic effect.

By the 1830s the native vegetation appears to have gone from the city area, except in parts of the Government Domain, behind the Rum Hospital and on Mrs Macquaries Point. Of these in 1879 the Director of the Botanic Gardens, Charles Moore, reported: 'The Eucalypts and other native trees in the Domain are fast dying out and will soon disappear altogether in that part toward the city; but on the eastern side very many of these are still in good health and will in all probability survive for many

The Vaucluse Bay shoreline is probably similar, though on a smaller scale, to the original shoreline of Sydney Cove. (1987)

'A place of wild flowers': the Outer Domain in the 1870s—the track winds out to Mrs Macquarie's Chair past trees of perhaps Bangalay, *Eucalyptus botryoides*. (Macleay Museum)

Old trees of Forest Red Gum, *Eucalyptus tereticornis*, in the Royal Botanic Gardens and Domain are descendants of the original woodland trees. (1983)

years'[2]. In his 1902 Annual Report[3], Director Joseph Maiden listed about 100 native plants 'growing without cultivation in the Outer Domain' with the comment: 'People are alive who remember it to be a place of wild flowers, and they have spoken to people who remember it in much the same state as it was when the white men first set foot on its shores'. Almost all of Maiden's plants were to disappear later in the typical clean-up and tidying programs that are incompatible with the growth and survival of natural populations of plants.

Today in the Royal Botanic Gardens and on Mrs Macquaries Point a few old warriors, the direct descendants of the 1788 trees, remain — Forest Red Gums, Port Jackson Figs, a couple of Blackbutts and Swamp Oaks, the occasional ageing Blueberry Ash and Cheese Tree, together with shrubs of *Clerodendrum tomentosum* and *Rapanea variabilis* that have survived amongst the rock outcrops. Efforts are being made to reinstate this vegetation along the foreshores of Woolloomooloo Bay.

Old Blackbutt trees, *Eucalyptus pilularis*, part of the Turpentine-Ironbark Forest, remain in Ashfield Park. (1988)

2 Ashfield

When the first Europeans arrived most of Ashfield was heavily timbered with open eucalypt forest growing on the clay soils developed from the Wianmatta Shale geology. Today the municipality is all built up, but the sort of trees there were originally can be deduced. If we look carefully we can even find a few of the original trees, or at least their direct descendants, still here in Ashfield. There are some old Turpentines, *Syncarpia glomulifera,* and Blackbutts, *Eucalyptus pilularis*, in Ashfield Park and in Albert Parade. Victoria Square still has a couple of Blackbutts and Red Mahoganies, *Eucalyptus resinifera,* and an old Grey Gum, *Eucalyptus punctata,* survives in the grounds of Summer Hill Public School. From this we can see that the forest of Ashfield was part of the Turpentine-Ironbark Forest that grew on the clay soils developed from the underlying Wianamatta Shale geology that covered much of the Inner West. For the smaller shrubs and herbs we can look for similar places where bush still survives to get some idea of the species. For instance bush on the same sort of soil in Concord West has shrubs of *Dodonaea triquetra, Melaleuca nodosa* and *Pittosporum undulatum*. Scattered through the forest would have been more open areas of *Themeda australis*, the Kangaroo Grass, herbaceous plants such as *Dichondra repens* and creepers such as *Hardenbergia violacea*, the Native Sarsaparilla.

On the swampy land at the head of Iron Cove and along Long Cove Creek (now the Hawthorne Canal at Haberfield) were mangroves and Swamp Oak or *Casuarina glauca* forest. These were filled and made into parks. There were Swamp Mahogany trees, *Eucalyptus robusta*, on the lower hill slopes, and some can still be seen in Robson Park. Nearby in the 1860s 'Ramsay's Bush covered all the area between the main [Parramatta] road and the waters of Iron Cove, and was a slice of primeval forest containing trees the like of which were rarely seen near Sydney. It was a favourite place in my boyhood days because of the native currants [*Leptomeria acida*] grown there,' reminisced a local resident[4].

Most of the eucalypt forest was cleared in the late nineteenth century, during a major period of suburban expansion. The *Railway Guide of New South Wales* gives us a glimpse of the Ashfield countryside in 1879, just before the housing boom:

> The course of the train [coming from Petersham] brings the tourist next, somewhat abruptly, by a viaduct over Long Cove Creek, a stream which flows along the bottom of the gorge, down which — away towards the Parramatta River — is suddenly disclosed a long vista of picturesque woods. The slender spire of St David's Presbyterian Church is seen amongst the trees to the north-west in the mid-distance. Away to the left are woods much nearer to the Line, and then the southern edge of the old Ashfield Racecourse is gained, with the old Southern Road from Sydney [Liverpool Road] on the right hand.[5]

The woods 'to the left' were soon to disappear under the suburbs of Lewisham and Summer Hill, while the trees around St David's, Ramsay's Bush, were to be replaced early in the twentieth century by the Brush Box-lined streets of the 'garden suburb' of Haberfield.

3 Auburn

Auburn is a small municipality located almost entirely on gentle Wianamatta Shale country. Unlike the suburbs closer in to Sydney where much of the forest had been cleared before the suburban expansion of the late nineteenth century, Auburn was not really developed for suburbs until after World War I.

Rainfall and temperatures experienced here are those of an area with mild wet winters and warm wet summers. Rainfall is sufficiently high to have supported open-forest, yet sufficiently variable, given the nature of the soils, for tree species from the drier Cumberland Plain, such as Grey Box, *Eucalyptus moluccana,* Broad-leaved Ironbark, *Eucalyptus fibrosa,* Stringybark, *Eucalyptus eugenioides,* Woollybutt, *Eucalyptus longifolia,* and Drooping Red Gum, *Eucalyptus parramattensis* to mingle with others from wetter areas such as Red Mahogany, *Eucalyptus resinifera* and Turpentine, *Syncarpia glomulifera.* The vegetation was much more scrubby than the more typical Turpentine-Ironbark Forest further east. Shrubs ranged from plants such as *Kunzea ambigua* and *Pultenaea villosa* that grow well in open, somewhat exposed positions, to *Glochidion ferdinandi, Breynia oblongifolia* and *Notelaea longifolia* that are normally plants of shaded forests.

A characteristic of remnants of the vegetation today is the abundance of paperbarks, *Melaleuca decora* and *Melaleuca nodosa,* sometimes called Tea-tree. Regrowth of *Melaleuca* may have been encouraged by the removal of the larger eucalypts, or by changes in the rate of burning, though its abundance is probably due to the poorly drained clay soils. It was conspicuous in 1879 for the *Railway Guide of New South Wales* notes: 'Leaving Homebush, [travelling west] the Railroad passes through an uninteresting piece of bush country, in which the (so-called) Tea Tree Scrub seems to be the principal feature'. Further on at Auburn, 'there is little or nothing to see. There is some sloping ground a considerable way to the left; a deep cutting or two, low scrubby trees with brushwood, and then a barren country'[5].

After crossing Duck River, 'An uninteresting barren bit of country follows, and then the Parramatta Junction is reached'. North of here the country is 'pleasing to the eye — wearied, as it naturally is, by the monotonous wilderness recently traversed by the train after passing Homebush'. In short Auburn's bush was evidently not of the calibre to impress the train traveller of the Victorian period.

Today our landscape has changed and, amidst a 'monotonous wilderness' of houses, we are lucky if we see remnants of this bush. There is still an important area on the western side of Duck River (in Parramatta) but in Auburn the best remnant of forest is at Silverwater, on land used for storing naval munitions. Here in 20 ha of forest are about 100 native species. Most common of the ten or so species of trees are *Syncarpia glomulifera, Eucalyptus paniculata, Eucalyptus fibrosa, Eucalyptus haemastoma* and *Eucalyptus globoidea.* The understorey is patchy with grassy open areas of *Themeda australis* and patches of shrubs of *Dodonaea triquetra, Kunzea ambigua, Zieria smithii* and *Polyscias sambucifolia. Dillwynia parvifolia, Eucalyptus*

Melaleuca decora and *Eucalyptus longifolia* (background) at the corner of Parramatta and Hill Roads, Auburn, in 1975, the last remnant of natural vegetation along the first road in the colony.

Small pockets of native scrub and grassland still survive in open areas and around graves in Rookwood Cemetery. Such areas may contain the only populations of naturally occurring native species for many kilometres, and it is important to protect them for their biological and landscape value. (J. Plaza, 1990)

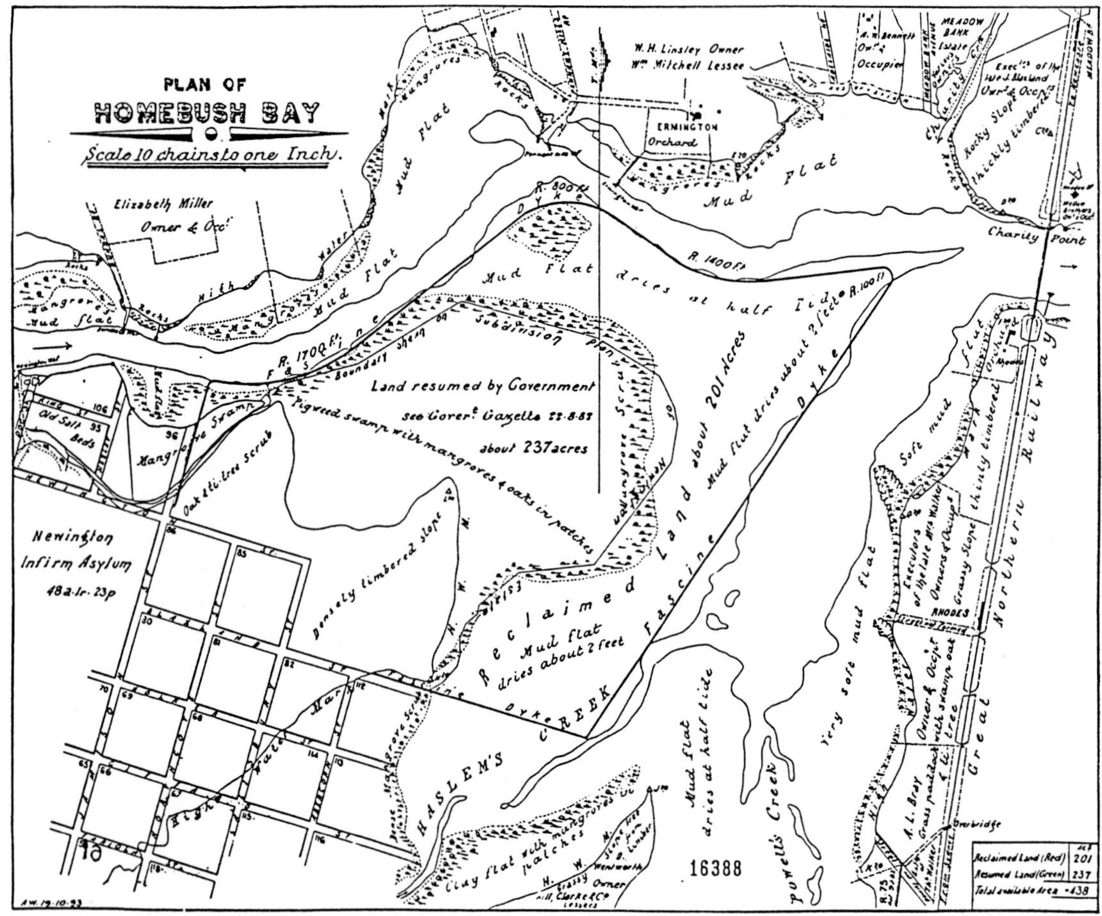

Homebush Bay in 1893, showing the extensive mudflats, mangrove scrub and 'pigweed' (presumably saltmarsh) swamp. At that time much of the area was proposed for reclamation.

In the 1950s extensive areas of Homebush Bay's saltmarsh and mudflat were destroyed by landfill operations. (Government Printing Office Collection, State Library of New South Wales)

longifolia and a small-fruited form of *Eucalyptus punctata* are species originally more common in the Inner West, but now quite rare. Along the Parramatta River near the Silverwater forest, is some of the best saltmarsh in Port Jackson, with populations of the rare *Halosarcia pergranulata* subspecies *pergranulata* and *Wilsonia backhousei*[6]. There is very little undisturbed saltmarsh in the nearby Bicentennial Park though it has extensive stands of the mangrove *Avicennia marina*. The permanent protection of the vegetation on the Navy land at Silverwater is needed. The forest-wetland vegetation forms a large clearly defined complex that should be managed as a nature reserve or as part of a 'Greater Bicentennial Park'.

There are also small bushland remnants in Rookwood Cemetery, that was described in 1861 as covered with 'dense ti-tree and wattle scrub and wooded with mahogany, stringybark, woollybutt and ti-tree'[7]. These remnants include the rare shrubs *Dillwynia parvifolia* and *Acacia pubescens,* and an unusual stand of Scribbly Gum, *Eucalyptus sclerophylla* with the shrubs *Banksia spinulosa, Angophora bakeri* and *Melaleuca erubescens.* Regular burning has been used to maintain parts of the cemetery and here Kangaroo Grass, *Themeda australis* and a variety of ground orchids can be found.

These small areas should be valued and not be cleared for grave sites or sprayed with herbicides. Local populations of wild plants are storehouses of the genetic diversity of populations once more common in the inner western Sydney suburbs. Rookwood is big enough for multiple use; its natural heritage is part of the cultural value of this marvellous Victorian necropolis.

4 Bankstown

Bankstown is the largest and most western of the inner western municipalities, and stretches from Auburn south to the Georges River. Like most of the others it is predominantly a landscape of gently undulating Wianamatta Shale soils, most of which has now been cleared of its native vegetation.

In the north soils are on the Bringelly Shale subunit and in the south from the Ashfield Shale subunit of the Wianamatta Shale. The vegetation appears to have been mainly woodland of *Eucalyptus moluccana* and *Eucalyptus tereticornis* with a grassy understorey and thickets of the shrub *Bursaria spinosa,* Blackthorn. Most of the shale areas have been cleared of native vegetation but important remnants remain, for example in Lansdowne Park and the Crest of Bankstown. Lansdowne Park is particularly important as it contains a very extensive area of this community with a diverse range of understorey species, including the now rare *Leichhardtia leptophylla* (this is probably the only known population in the Sydney district), *Chorizema parviflorum, Pimelea spicata* and *Zornia dyctiocarpa.* This vegetation was similar to that of the Cumberland Plain Woodlands further west. Some interesting rainforest-type species still occur in sheltered gullies near Marion Street on the small escarpment between Milperra and Punchbowl. South of here appears to have been open-forest of Turpentine, *Syncarpia glomulifera* and Broad-leaved Ironbark, *Eucalyptus fibrosa.* Here the Turpentine-Ironbark Forest reaches its western limit. Norfolk Reserve at Greenacre contains an interesting remnant with Woollybutt, *Eucalyptus longifolia,* and Ironbarks, *Eucalyptus crebra* and *Eucalyptus fibrosa,* also part of this transitional vegetation. This vegetation has almost completely gone from Bankstown.

Perhaps the most interesting vegetation on the shale areas is found in the broad valleys of creeks such as upper Salt Pan Creek and Duck River, in the suburbs of Leightonfield, Birrong, Regents Park and Bankstown itself. Here the trees were mainly *Eucalyptus fibrosa-Eucalyptus moluccana,* Ironbark-Grey Box, but with a very distinctive paperbark shrub layer of *Melaleuca decora* and *Melaleuca nodosa.* Also here were trees of *Eucalyptus longifolia, Eucalyptus sideroxylon, Eucalyptus parramattensis* and *Eucalyptus sclerophylla,* and a variety of small shrubs including *Pultenaea villosa, Hakea sericea, Kunzea ambigua, Acacia falcata, Epacris purpurascens,* green-flowered *Callistemon pinifolius* in damp places, and sometimes the now rare *Pultenaea pedunculata, Persoonia nutans* and *Acacia pubescens.* Examples of this vegetation are still found in Carysfield Park at Bass Hill, Leightonfield and Condell Park. Small populations of *Acacia pubescens* may be seen along the railway line, on bare clay embankments, between Punchbowl and Regents Park.

Natural vegetation is also still found along parts of the Georges River, particularly in flood-prone sites unsuitable for residential development. At Deepwater Park, at East Hills, is a remnant of the River-flat Forest with a number of Blue Box trees, *Eucalyptus bauerana.* Unfortunately most of the original understorey that grew with these trees has been removed to make cleared areas for picnicking and car parking, when more open areas nearby could have been used. Nearby patches of *Eucalyptus eugenioides-Eucalyptus fibrosa* woodland show how much diversity and interest the understorey plants can add

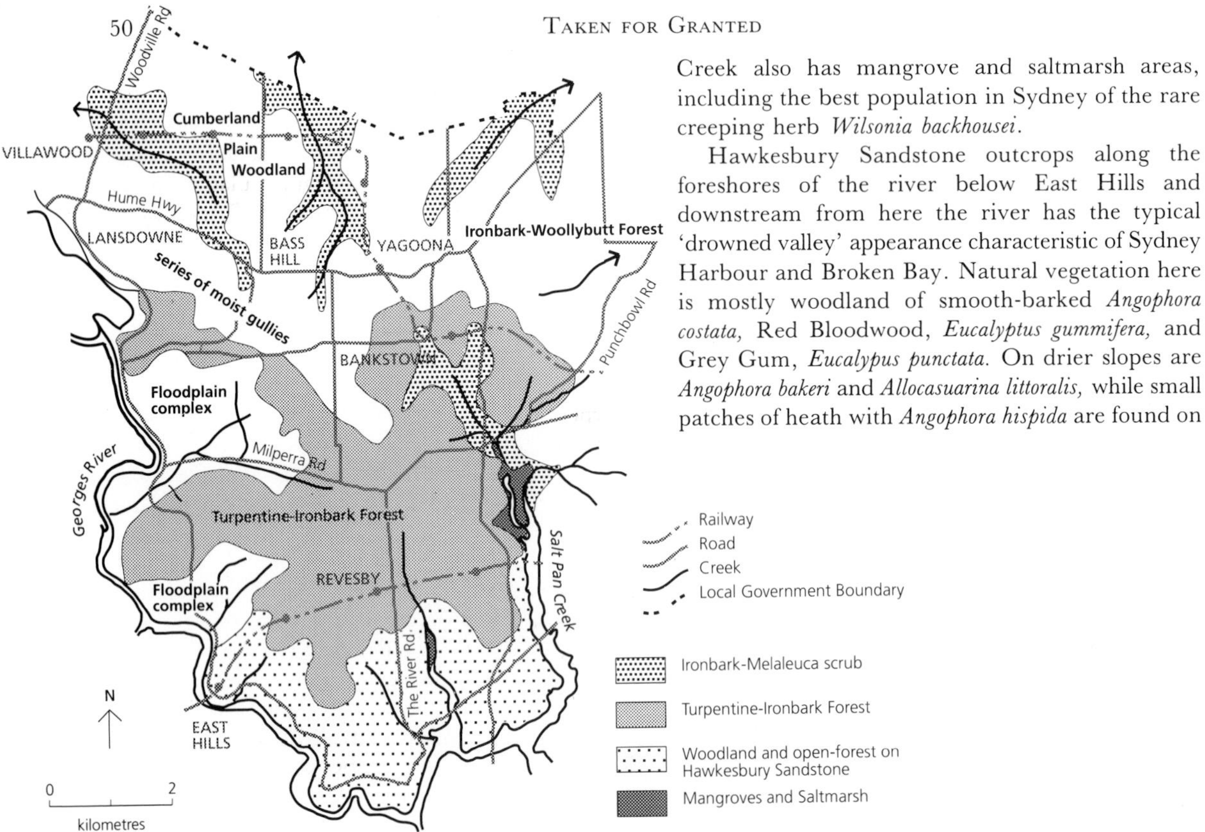

Probable nature and extent of original vegetation in Bankstown.

to recreation areas. Here are small trees of *Backhousia myrtifolia*, used by Aborigines to make boomerangs, *Polyscias sambucifolia* with small flat purple fruits, and *Notelaea* with large olive-like ones. Crimson bells of *Correa reflexa*, bright orange *Pittosporum revolutum* fruits, and white sprays of *Rulingia pannosa* flowers are interspersed with feathery-leaved *Acacia decurrens, Dodonaea multijuga* and *Pultenaea villosa* shrubs, and occasional patches of soft *Adiantum aethiopicum,* Maidenhair Fern. There are also some very interesting floodplain thickets with *Melaleuca ericifolia* and *Melaleuca linariifolia* and patches of sedgeland accessible from well constructed boardwalks. Also along the Georges River are strips of mangroves, *Avicennia marina* and *Aegiceras corniculatum,* and small patches of Swamp Oak, *Casuarina glauca* forest. Further downstream between Picnic Point and Lugarno, the alluvial flats have been developed as picnic areas, leaving the trees, but carefully clearing the understorey and replacing it with mown grass. Whilst such areas may be needed for public recreation, similar development of all such river flats is destroying the last populations of the small shrubs and herbaceous species native to these areas. Salt Pan Creek also has mangrove and saltmarsh areas, including the best population in Sydney of the rare creeping herb *Wilsonia backhousei*.

Hawkesbury Sandstone outcrops along the foreshores of the river below East Hills and downstream from here the river has the typical 'drowned valley' appearance characteristic of Sydney Harbour and Broken Bay. Natural vegetation here is mostly woodland of smooth-barked *Angophora costata,* Red Bloodwood, *Eucalyptus gummifera,* and Grey Gum, *Eucalypus punctata.* On drier slopes are *Angophora bakeri* and *Allocasuarina littoralis,* while small patches of heath with *Angophora hispida* are found on some of the gravelly ridge-tops. These sandstone areas were the last part of Bankstown to be developed and, with foresight, some of these hillsides have been included in Georges River State Recreation Area.

In Deverall Park at Condell Park a small population of *Epacris purpurascens* in a fragment of diverse shrubby woodland has recently been saved from destruction. (1988)

Rare shrubs manage to survive in unmown 'islands' of Ironbark and Paperbark trees in Carysfield Park, Bass Hill, but for how long? Whittling away of bushland by mowing poses a threat in many bushland parks. (1989)

Wetland zonation can be seen from boardwalks in Deepwater Park, Milperra. Here reedy *Phragmites australis*, shrubs of *Melaleuca erubescens* and *Casuarina glauca* trees can be seen in sequence. (1988)

Unfortunately most of the understorey has been removed from beneath the *Eucalyptus bauerana* trees that formed River-flat Forest on the rich alluvial soil of Deepwater Park. (1984)

Burwood Villa, a watercolour by Joseph Lycett (1774–1828), painted around 1824. Lycett's accompanying text asserts the 'Forest Scenery is characteristic of the country', and notes 'how speedily the Forest in New South Wales can be cleared of its superfluous timber, and rendered contributable to the comforts and luxuries of man.' (Art Gallery of South Australia, Adelaide, M.J.M. Carter Collection 1988)

5 Burwood

Burwood is a small municipality located almost entirely on the clayey soils formed on the Wianamatta Shale. Clearing the vegetation for grazing lands began early in the nineteenth century. Today there is virtually nothing left of its original vegetation of Turpentine–Ironbark Forest, and Joseph Lycett's description of Burwood Villa in 1824 perhaps provides the explanation:

This Estate is within eight miles of Sydney, on the high road to PARRAMATTA, and bounded at the back by the high road to LIVERPOOL, comprising a square of one thousand acres, within a rail-fence; and is a remarkable instance how speedily the forest in NEW SOUTH WALES can be cleared of its superfluous timber, and rendered contributable to the comforts and luxuries of man; for, within three years of the felling of the first tree on this estate, the whole was enclosed and subdivided, five hundred acres were more or less cleared; a desirable Villa-House, with every convenient Appendage, was erected; artificial Grasses were growing, in aid of the natural Pasture; and a Garden of four acres was in full cultivation, containing upwards of three hundred Trees.[8]

In December 1834 the *Sydney* [*Morning*] *Herald* advertised allotments of the 'Burwood' estate with 'timber: Shingle Oak, Iron and Stringy Bark, Mahogany, Blue Gum etc.'. These were probably trees of *Allocasuarina torulosa, Eucalyptus paniculata, Eucalyptus globoidea, Eucalyptus resinifera* and *Eucalyptus tereticornis* (or perhaps *Eucalyptus saligna*) respectively.

Large residences were built later as the area became a fashionable semi-rural retreat from the city. A picture of the landscape around Croydon and Burwood at this time is given in the *Railway Guide of New South Wales* for 1879.

The railway passenger, on leaving Ashfield Station, is now (for about a mile) hurried past an agreeable bit of home scenery; diversified by gardens and trees, with a wide, uneven space on either side of the road in the background, where Nature has not yet been ruthlessly *improved* away. Streets (for the most part mere lanes) intersect this tract, whereon stand villas and gardens belonging to Sydney people, displaying a considerable amount of domestic comfort, originality, and even elegance of design. Vistas of pleasant country roadways — green, and as yet innocent of dust and mire — stretch up the gentle eminences to the left and right; and then the sombre eucalyptus, intervening market gardens, and rural homesteads, successively meet the eye before the train reaches Croydon Platform. Here the train (if suburban) may possibly stop for a moment, but probably speeds onward unchecked to Burwood.[5]

Higher density suburban development had begun by 1900 and by 1930 the last vestiges of natural vegetation had gone. A small volcanic outcrop between Livingstone Street and Woodside Avenue, Burwood, had been quarried by 1866. It may have had a slightly different flora.

6 Canterbury

Canterbury is a roughly triangular municipality between the Cooks River and its tributary Wolli Creek, flowing to Botany Bay, and Salt Pan Creek to the west, flowing to the Georges River. Most of this country is gently undulating, with clay soils developed from Wianamatta Shale, and would have once been covered by Turpentine–Ironbark Forest. Ironbarks appear to have been particularly common. There are remaining trees of *Eucalyptus fibrosa* in Wiley Park and Wiley Park Girls High School, and old records suggest that Belmore was near the southern limit of *Eucalyptus siderophloia,* the Northern Grey Ironbark. The forest was destroyed by the surge of suburban development that followed the opening of the Bankstown railway line, beginning at Hurlstone Park and Canterbury in the 1890s, and moving westwards through Campsie, Belmore, Lakemba and Punchbowl in the 1920s and '30s.

At the eastern end of the municipality, in Earlwood, Hurlstone Park and Canterbury, Hawkesbury Sandstone outcrops. This provides a more rugged landscape with steep hillsides and cliffs. The vegetation here was woodland with trees of the smooth-barked *Angophora costata,* Blackbutt, *Eucalyptus pilularis,* Sydney Peppermint, *Eucalyptus piperita,* Red Bloodwood, *Eucalyptus gummifera,* and Turpentine, *Syncarpia glomulifera,* with a varied shrub understorey. Remnants of this woodland vegetation can be seen along Wolli Creek at Girrahween Park, the best natural area in Canterbury Municipality, and Nannygoat Hill at Earlwood. Shrubby heath can be seen at Highcliff Road, Undercliffe. The proposed F5 Freeway along Wolli Creek Valley will destroy the amenity of these last valuable bushland areas. Sheltered sandstone cliffs and railway cuttings may have native species particularly ferns, *Gleichenia dicarpa, Culcita dubia* and *Pteridium esculentum,* the Kangaroo Grass *Themeda australis* and the shrub *Kunzea ambigua,* and there are some lovely Blackbutts on the steeper slopes of Earlwood.

The Cooks River was explored in December 1789 by Lieutenant William Bradley of the *Sirius*: 'I found it to be a Creek of about 8 miles length to the NW with a winding shoal channel & end in a drain to a swamp, all saltwater'[9]. There were extensive mangrove and saltmarsh flats in the estuary and a dam was constructed in 1840 at Tempe to provide fresh water. By 1870 the river had been polluted with sewage and rubbish, and the dam was silting up as a result of uncontrolled agricultural, industrial and suburban development in the catchment. Like the Tank Stream before it, the Cook's River provided an early example of pollution in Sydney's waterways. In 1895 there were alarms at the unsanitary conditions and the Tempe Dam was lowered. This caused only short term improvement, and in 1925 the Cooks River Improvement League was formed by environmentally conscious people. Pressure on the government led to the concreting of the upper reaches as Depression relief work and, with the 1946 Cooks River Improvement Act, the removal of the tidal gates at Tempe, the dredging of the river and the filling of the low-lying wetlands for parklands. At the same time the mouth of the river was relocated to allow upgrading of Kingsford Smith Airport, and the extensive mangrove and saltmarsh flats were destroyed by landfill.

Re-opening the river to tidal flushing improved conditions for wildlife and in the 1940s clean sandy reaches with plentiful prawns are remembered by local residents. Since then increasing pollutants and silt in stormwater and industrial waste have again degraded the river. Today concrete or iron

The Cooks River at Undercliffe early this century still had natural banks with reeds and remnants of the original forest. (Mitchell Library, State Library of New South Wales)

Girrahween Park at Earlwood, one of the few bushland parks in Canterbury, has a surprisingly rich variety of shrub species. Here, *Platysace lanceolata, Acacia longifolia, Persoonia levis* and *Kunzea ambigua* can be seen growing amongst *Angophora costata* and *Eucalyptus gummifera*. (J. Plaza, RBG, 1990)

embankments have replaced the natural riverbank, and the adjacent floodplain has been filled, but short sections of the river at Canterbury still have fringing banks of *Phragmites australis,* Common Reed, with occasional recolonising Grey Mangroves, *Avicennia marina*.

Along the river valley were forests of *Casuarina glauca*, Swamp Oak, and *Eucalyptus robusta,* Swamp Mahogany, *Eucalyptus tereticornis*, Forest Red Gum and possibly *Eucalyptus saligna*, Sydney Blue Gum. *Casuarina glauca* has been planted in the parks along the river but, for the long-term improvement of the valley, more of the larger growing eucalypts should be planted. These will provide much-needed native bird habitats and restore the natural vegetation structure.

Unlike the Cooks River, Salt Pan Creek, which runs down to the Georges River, is in reasonable natural condition and there are still good mangrove and some more limited saltmarsh areas at Riverwood, though their extent is threatened by landfill and freeway construction.

Scrub with *Banksia serrata* and *Kunzea ambigua* amongst the sandstone outcrops of Nanny Goat Hill overlooking Wolli Creek at Undercliffe. Part of a very narrow remnant of bushland along the valley, its amenity would be severely impaired by a proposed freeway. (J. Plaza, RBG, 1990)

7 Concord

Concord is a largely residential and partly industrial municipality, on an irregularly shaped peninsula, bounded by a series of small bays along the Parramatta River. The central part of Concord is undulating, with clay soils developed from Wianamatta Shale. When the first Europeans arrived most of this was heavily timbered with open eucalypt forest. A surveyor's report of 1857 described land on the eastern side of the present Concord Golf Course, 'good forest land...on good black soil...wooded with gum, blackbutt, stringybark, mahogany, apple, ironbark and she-oak'[10]. The nearby Thomas Walker Convalescent Hospital 'was a dense brush, the haunt of hundreds of flying-foxes' in the 1850s. These trees would have been *Eucalyptus tereticornis,* Forest Red Gum; *Eucalyptus pilularis,* Blackbutt; *Eucalyptus globoidea,* White Stingybark;

Eucalyptus resinifera, Red Mahogany; *Angophora floribunda,* Rough-barked Apple; *Eucalyptus paniculata,* Grey Ironbark; and *Allocasuarina torulosa*, Forest She-oak respectively. Together with *Eucalyptus punctata,* Grey Gum, *Eucalyptus moluccana*, Grey Box and *Syncarpia glomulifera*, Turpentine, these trees were part of the Turpentine–Ironbark Forest that covered much of the inner western suburbs.

At Concord West a small remnant of this forest has survived European occupation. In a corner of the magnificent grounds of the Dame Eadith Walker Hospital, 'Yaralla', Sydney's last great landed estate and as important as Melbourne's Rippon Lea, is the Yaralla Bush, with trees of Grey Ironbark, White Stringybark, Red Mahogany, Grey Gum, Turpentine and Rough-barked Apple with an understorey of about 70 native shrubs, herbs and grasses. A number of exotic species are present and bushland management work has been carried out by the National Trust.

Elsewhere old Turpentines and an occasional Blackbutt remain in Queen Elizabeth Park; in Concord Golf Course there are Grey Ironbarks along Majors Bay Road and a group of healthy Forest Red Gums on the Cumming Avenue side. There are still Rough-barked Apples and White Stringybarks in North Strathfield Public School grounds, though an unusual Northern Grey Ironbark, *Eucalyptus siderophloia,* was removed some years ago.

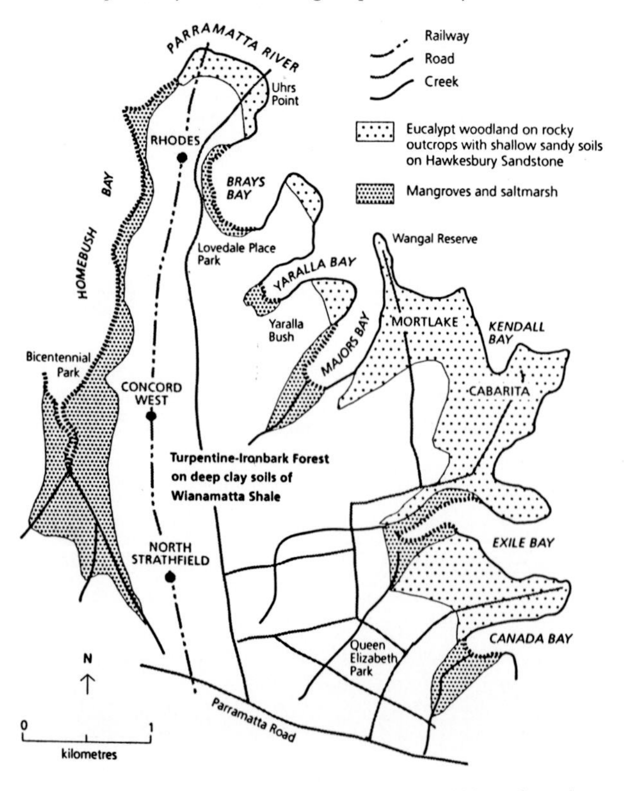

Vegetation patterns in Concord as they would have been in 1788.

On the Parramatta River headlands, from Rhodes Point to Bayview Park on the eastern side of Concord, the Wianamatta Shale has eroded exposing the underlying Hawkesbury Sandstone. The soils are shallow and sandy and the original vegetation was different from that on the shale, a woodland of small trees with a dense shrubby undergrowth. In some places — Cabarita Park, Bayview Park and Prince Edward Park — there were sandstone rock platforms with patches of *Kunzea ambigua* heath and clumps of *Lomandra longifolia.* The trees on the headlands included Blackbutt, Red Bloodwood, *Eucalyptus gummifera,* and Smooth-barked Apple, *Angophora costata.* Among the shrubs were Coast Banksia, *Banksia integrifolia,* Black She-oak, *Allocasuarina littoralis,* Cheese Tree, *Glochidion ferdinandi,* and *Melaleuca nodosa.* Remnants of this kind of vegetation can be seen in Putney Park across the Parramatta River in Ryde, though it has disappeared from Concord. Recent landscape plantings at Prince Edward Park and Wangal Reserve have included local indigenous native species.

During the nineteenth and early twentieth centuries, extensive mangrove and saltmarsh flats were conspicuous at the heads of the bays and inlets of the Parramatta River. Extensive landfill reclamation programs, beginning in the 1920s, continued to the 1970s and successfully eliminated the saltmarsh flats and most of the associated mangroves. Canada Bay (now St Lukes and Cintra Parks and residential land), Exile Bay (now Massey, Edwards and Greenlees Parks), Majors Bay, Yaralla Bay and Brays Bay (now McIlwaine Park) were all sites where wetlands were used as garbage tips. Fringes of mangroves are still to be found in Brays, Yaralla and Majors Bays, and smaller pockets in Kendall and Exile Bays, and at Uhrs Point. A rare surviving example of mangrove, saltmarsh, mud flat and Swamp Oak zonation can be seen adjacent to the Concord Hospital parking area and Lovedale Place Park. There is also an unusual occurrence of the paperbark tree *Melaleuca styphelioides,* and the most easterly occurrence of Grey Box, here. In Homebush Bay, on the western side of the municipality, and now part of Bicentennial Park, are extensive tidal flats with an abundance of *Avicennia marina*, the Grey Mangrove, though the extensive saltmarsh and mudflats once here were filled in the 1960s.

Trees of Blackbutt, *Eucalyptus pilularis*, 30 m high (left) and Swamp Mahogany, *Eucalyptus robusta*, 15 m high (centre), photographed in Queen Elizabeth Park, formerly Concord Park, Concord early in the century. There are still a considerable number of remnant trees, mostly Turpentines, *Syncarpia glomulifera* in the park. A large Blackbutt, *Eucalyptus pilularis*, shown here (right), survives near the war memorial. (J. Plaza, RBG, 1990)

Most of the saltmarsh of Concord has been filled for parkland. This remnant beside Lovedale Place Park near Concord Hospital is particularly important, as it is one of the few remaining sites along the Parramatta River where the zonation of mangroves, saltmarsh, mudflat, *Juncus* meadow and *Casuarina glauca* swamp forest can still be seen. (J. Plaza, RBG, 1990)

8 Drummoyne

Drummoyne occupies a broad, relatively low, Hawkesbury Sandstone peninsula on the southern side of the Parramatta River. Soils here are sandy loams, and the original vegetation, now completely replaced by suburban housing, would have been forest and woodland, typical of sandstone slopes. Characteristic trees were *Eucalyptus pilularis,* the Blackbutt; the smooth-barked *Angophora costata;* and *Eucalyptus piperita,* the Sydney Peppermint. Large spreading Port Jackson Figs, *Ficus rubiginosa,* would have been conspicuous on the rocky headlands. Mangroves and saltmarsh flats occurred in the intertidal zone in Iron Cove and Hen and Chicken Bays, but have been filled to provide parks. Odd plants of *Avicennia marina,* the Grey Mangrove, and *Sarcocornia quinqueflora,* Samphire, may still be seen near Henley Marine Drive. Mary Salmon evoked the scene at Vault Point in 1904, 'a wild isolated point of rock and mangrove swamp land between the Iron Cove and Long Cove, where the wild cherry [possibly *Syzygium paniculatum*], the she oak and the gum tree grow in native abandon'[11]. On rocky outcrops here a few clumps of native sandstone ground cover species, in particular, *Lomandra longifolia*, still grow.

Clumps of *Lomandra longifolia* growing on sandstone outcrops at Vault Point are among the few surviving representatives of the original vegetation of Drummoyne. (1988)

Five Dock, in the south-western corner, is on the edge of the Wianamatta Shale mantle that covers most of the inner western suburbs. 'The roads are through large estates, many of them covered with a

Deteriorating forest remnant on the sandstone foreshores at Abbotsford in the 1920s. The trees include Blackbutt, *Eucalyptus pilularis*. (Mitchell Library, State Library of New South Wales)

dense growth of forest timber' reported *The Echo* in 1890[11]. Turpentine-Ironbark Forest similar to that of adjoining Ashfield and Concord municipalities would have grown here.

9 Hurstville

Hurstville is a south-western municipality bounded by the Georges River, Salt Pan Creek, and the East Hills and Southern railway lines.

The northern half of the municipality is Wianamatta Shale country and would have once had Turpentine-Ironbark Forest. In the nineteenth century, 'Lords Forest' and 'Connells Bush', now Mortdale, Peakhurst and Hurstville, provided timber by way of the 'Forest Road'. With the coming of the Southern Railway (1884), these areas developed as suburbs in the early twentieth century. The East Hills Line (1931) brought similar development to the southwest. Almost all the shale vegetation has gone. There is a small remnant in Riverwood Park[12], and an important group of trees including *Eucalyptus paniculata, Eucalyptus punctata, Syncarpia glomulifera, Angophora floribunda* and, in particular, *Eucalyptus longifolia* in Olds Park at Penshurst. Additional planting of trees grown from local seed will help maintain this remnant of the forest of Forest Road. A group of *Syncarpia* trees survive in River Road, near the railway at Oatley.

Outcrops of Hawkesbury Sandstone along the foreshores and headlands of the Georges River feature most of Hurstville's remaining native vegetation. This is mainly woodland of smooth-barked *Angophora costata*, Red Bloodwood, *Eucalyptus gummifera,* and the Sydney Peppermint, *Eucalyptus piperita*. The 45 ha Oatley Park is by far the largest intact sample of Hawkesbury Sandstone vegetation in the St George district and its bushland is generally healthy and relatively weed-free because of its position — no creeks from urban areas drain into it and there is only a short boundary in contact with housing. Other examples are in Oatley Heights Park, the Georges River S.R.A. parklands, H.V. Evatt Park and Boggywell Creek, all at Lugarno, the southern end of Hurstville Golf Course and Yarran Road Reserve at Oatley. H.V. Evatt Park is interesting for an area of volcanic soil, much more fertile than the surrounding sandstone, that would probably have had 'rainforest' species. Though cleared early last century for an orchard, there are still some Coachwoods, *Ceratopetalum apetalum*, at the far western end[12].

Along Salt Pan Creek, mangroves and saltmarsh are found south of Henry Lawson Drive. At Edith Bay along the Georges River is an extensive stand of the paperbark, *Melaleuca linariifolia.* Swamp Oak, *Casuarina glauca,* forest is common in some of these reserves.

Como Railway Bridge from Neverfail Point about the turn of the century showing *Angophora costata* trees on the sandstone hillsides of the Georges River. Beyond the bridge is an island of mangroves. (Macleay Museum)

Road construction near Kyle Bay in 1933. An extensive area ot sandstone scrub and woodland is being made accessible for housing. (Government Printing Office Collection, State Library of New South Wales)

10 Kogarah

Kogarah Municipality occupies the northern foreshores of the Georges River from Como Railway Bridge to the Captain Cook Bridge at Sans Souci, and extends northwards to the railway line at Kogarah. Most of this area is on Hawkesbury Sandstone country, though Wianamatta Shale extends from adjacent Hurstville across the railway line between Allawah and Mortdale. The clay soils here would have once had Turpentine-Ironbark Forest but this was logged and cleared in the nineteenth century. The poor sandy soils on the sandstone lands were unsuitable for agriculture, but unlike other municipalities where the generally steeper topography allowed some natural vegetation to remain, in Kogarah, very little bushland survived the extensive suburban development of the first half of this century. Almost all, including most foreshore vegetation, has gone.

The best remnants are at Oatley Point and Carss Bush Park. At Carss Bush Park there is still forest of Blackbutt, *Eucalyptus pilularis,* and a variety of

Carss Bush Park at Blakehurst is one of the few parks in Kogarah still retaining some natural bush. Here a large Blackbutt grows amongst the sandstone outcrops with shrubs of *Banksia* and *Platysace*. (J. Plaza, RBG 1990)

smaller shrubs, though encroachment by mowing along the edges, and planting of exotic plants, is causing deterioration. At the southern end of Carss Bush Park in a swampy site is the only stand of Swamp Mahogany, *Eucalyptus robusta*, in the St George area.

Grey Ironbark and Blackbutt occur at Oatley Point Reserve, adjoining the foreshore of Oatley Bay. This appears to be the only occurrence of Grey Ironbark along the foreshores of the Georges River.

11 Leichhardt

The original forests of Leichhardt were cleared early in the nineteenth century for the agricultural estates of colonial identities such as Major George Johnston of 'Annandale' and Surgeon John Harris of 'Ultimo'. By 1900, subdivision for suburban allotments and the construction of rows of Victorian terraces and villas had virtually removed all of the native vegetation, leaving few traces of its original appearance. However, we may surmise that on the clay soils derived from Wianamatta Shale which largely make up the suburbs of Leichhardt, Annandale and Lilyfield, there would have been Turpentine-Ironbark Forest. Peter Cunningham described the land along Parramatta Road between Annandale and Ashfield in 1827:

> On each side of the road is a post and rail fence, while the land is thickly covered with heavy timber and brush, the soil being usually a poor shallow reddish or ironstone clay, the contemplation whereof presents but little pleasure to the agriculturalist.[13]

Trees here would have included the Grey Ironbark, *Eucalyptus paniculata,* Red Mahogany, *Eucalyptus resinifera,* and Blackbutt, *Eucalyptus pilularis.*

On the more rugged Hawkesbury Sandstone landforms — the harbourside suburbs of Glebe and Balmain — would have been typical Sydney sandstone open-forest, with trees of smooth-barked *Angophora costata* and Sydney Peppermint, *Eucalypus piperita.* The species present would have been similar to bushland found today on the nearby northern side of the harbour, such as at Balls Head and Berry Island. A few tough plants of *Lomandra longifolia* still persist on rock outcrops at Callan Park. Blackwattle Bay appears to take its name from the small tree *Callicoma serratifolia* that probably occurred along creek lines entering the bays.

This 1859 pencil sketch by Conrad Martens captures the nature of the steep wooded slopes and overhanging sandstone rock ledges that were to disappear during the next 40 years under the terrace houses of Glebe. (Mitchell Library, State Library of New South Wales)

Forest trees surround the newly established Callan Park Hospital on Iron Cove. (Macleay Museum)

12 Marrickville

Marrickville Municipality largely occupies undulating country with clay soils developed from Wianamatta Shale. Originally clothed with Turpentine–Ironbark Forest, most was cleared for agriculture early in the nineteenth century. In 1830 the *Sydney Gazette* reported:

> The timber on the Petersham estate (the property of Dr Wardell) has been valued at the extra ordinary sum of £40,000. It comprises two or three thousand acres, and is in some places very thickly wooded. Incredible as the estimate may appear, we are inclined from the increasing scarcity and dearness of firewood, to think it not much above the mark.

There was evidently a general shortage of timber as a result of clearing, and the fuel demands of the growing town of Sydney. By 1879 the suburbs were expanding, and bushland remnants that had survived on the agricultural estates were destroyed in the late nineteenth century housing boom.

At the southern end of the municipality, the Cooks River follows a narrow valley with steep Hawkesbury Sandstone slopes that contrast with the gentle shale landform elsewhere. Here there would have been Blackbutt, *Eucalyptus pilularis*, forest with typical sclerophyllous understorey shrubs, remnants of which may still be seen along Wolli Creek, in particular in Girrahween Park in Canterbury Municipality. There are still some remaining Blackbutts in Marrickville Golf Course, and to enhance the river, some of the sandstone slopes here could be effectively landscaped with local indigenous species.

Extensive mangrove and saltmarsh flats, along the Cooks River as far up as Illawarra Road, were filled for playing fields after World War II. *Casuarina glauca*, Swamp Oak, forest was also common along the river and there are still old trees on Marrickville Golf Course, as well as many younger plantings.

Residence in Harriet Street, Marrickville, in 1871 with a well established ornamental garden surrounded by eucalypt forest. (Mitchell Library, State Library of New South Wales)

The Cooks River flooding a substantial stand of Swamp Oaks, *Casuarina glauca*, in 1925. Remnant trees still survive along the river, though many additional plantings have been made. (Mitchell Library, State Library of New South Wales)

A marvellous view of the estuary of the Cooks River in the 1880s taken from 'The Warren', now Warren Park, Marrickville South. In the foreground, just outside the garden is Scribbly Gum woodland on the sandstone hillside. Beyond is Unwins Bridge and the Cooks River Dam. Extensive saltmarshes and mangroves are visible. (Mitchell Library, State Library of New South Wales)

13 Rockdale

Rockdale Municipality extends from the western shore of Botany Bay, south of Cooks River, to Kingsgrove. A geologically varied area, this would have had a variety of plant communities, though only very small remnants survive.

Behind Lady Robinsons Beach on Botany Bay, and up to a kilometre inland, were a series of north-south beach sand ridges and swamps. Trees of '*Eucalyptus pilularis, E. botryoides, E. robusta, Banksia integrifolia,* and *Angophora lanceolata* [= *costata*] are the common large plant growths of the sandy waste, while *Eucalyptus robusta* and *Casuarina glauca* are the common large growths of the swampy areas,' reported the geologist E.C. Andrews in 1912[14]. Of one of these swamps, Pat Moore's or Patmore Swamp, 'a fairly heavy growth of bang alleys [*Eucalyptus botryoides*] and swamp mahogany trees [*Eucalyptus robusta*] abutted on the western side of this swamp — notable trees in every way, either in girth or foliage they were, and no undergrowth, but a fine coating of native grasses of a meadow character,' was how a local resident recalled them as being in the 1870s[15]. An 1889 newspaper reported: 'The stretch of country between Kogarah and Sans Souci is pleasant to look upon. It is of goodly make. Wide plains of fern, clean stretches of trees and bright-flowered plants. The population along the route was as sparse as the passengers on the tram were'[16]. By draining and adding nightsoil, the swamps were converted to market gardens. In the 1920s and '30s the sand ridges were built on. Only one small example of the original forest remains, in the Leo Smith Reserve at Ramsgate. Here are the *Angophora costata* and *Eucalyptus botryoides* trees, as described by Andrews, with an understorey of shrubs including *Monotoca elliptica, Breynia oblongifolia, Banksia serrata, Acacia longifolia,* the fern *Pteridium esculentum,* and scramblers *Hibbertia scandens* and *Billardiera scandens.* Nearby on alluvium of Muddy Creek is swamp

At Undercliffe in 1919 the river banks had reeds of *Phragmites australis* and *Schoenoplectus litoralis* with trees of *Casuarina glauca*. Patches of *Phragmites* still grow along the river's edge at Canterbury, although the building of steel and concrete walls along the river in the 1940s destroyed most of the natural banks. (Linnean Society of New South Wales)

Arncliffe Station, a bushland valley in the 1880s where 'the native flowers of Australia are still to be found in profusion'. (Rockdale Municipal Library, Local History Collection)

woodland with *Casuarina glauca,* Swamp Oak, and *Eucalyptus robusta,* Swamp Mahogany. Exotic pines were planted years ago and the invasive weed species *Lantana camara* has established, but this last small example of original vegetation is very valuable.

Muddy Creek drains into the Cooks River where there were extensive mudflat, saltmarsh and mangrove flats. A small wetland in Barton Park near Eve Street is being rehabilitated by the Water Board. Estuarine wetlands followed Wolli Creek, as far as the weir at Turrella, where patches of mangrove and saltmarsh can still be seen. In the early nineteenth century, mangrove wood was burnt to supply soda for the manufacture of soap. In 1828 there were 'two or three Manufactories of Soap' at Botany Bay, though by 1831 it was reported that 'The soap boilers still suffer considerable restriction from the insufficient supply of mangrove ashes'[16]. Also along Wolli Creek (known in 1833 as Cabbage Tree Creek[17]), on the black alluvial soils, would have been a swamp forest of *Eucalyptus robusta* and *Casuarina glauca* with the paperbark *Melaleuca linariifolia,* Lillypillies, *Acmena smithii,* and presumably Cabbage Palms, *Livistona australis,* though no palms have survived here.

Arncliffe, Banksia, Bardwell Park, Rockdale, Kogarah and Sans Souci occupy steeper sandstone country, the nature of which gave the name to Rockdale. The bushland began to disappear under houses from the 1880s onward. 'ARNCLIFFE — a new suburb on the Illawarra Line where, in spite of the speculative builder and thousands of excursionists the native flowers of Australia are still to be found in profusion,' reported an 1889 guide book[16]. The builders won the day and only small remnants of bushland survive along Wolli and Bardwell Creeks.

Woodland of *Angophora costata* with a dense ground cover of Bracken and shrubs in the Leo Smith Reserve at Ramsgate is an important remnant of woodland on coastal sand. (J. Plaza, RBG, 1990)

Along the upper Bardwell Creek valley is open-woodland of *Angophora costata* and *Eucalyptus piperita,* with a shrubby understorey with up to 80 different species. Between Bexley Road and the Bardwell Valley Golf Course is shrubland of *Kunzea ambigua* and *Casuarina littoralis* and heath with *Epacris pulchella, Epacris longiflora, Epacris microphylla, Astroloma pinifolium* and *Styphelia tubiflora,* some of it on skeletal soil on an old scraped quarry site. The bared rock has kept the taller growing shrub species out, allowing the heath species, which are now rare in this part of Sydney, to prosper. Vegetation remaining along Wolli Creek is mainly on the Earlwood side, but Stotts Reserve at Bexley North has a rare remnant of *Eucalyptus saligna* open-forest along a side creek, as well as the more typical *Angophora costata-Eucalyptus piperita* woodland[18].

At the western end of Rockdale, the suburbs of Bexley and Kingsgrove are on the edge of the Wianamatta Shale. These areas probably had Turpentine-Ironbark Forest which was eagerly sought by the timber getters. In 1833, on the estate of 'Bexley' there were 'valuable quantities of timber. . . stringy and ironbark, black-butt, mahogany, shingle-oak, turpentine, red, blue and whitegum, honeysuckle for ship and boat builders, and white wood of a large size, so much used by coach-builders and others'[17]. These forests have disappeared completely, though some Turpentines in the upper part of the Bardwell Valley are probably indicative of enrichment by shale downwash.

The remnants of bushland in Rockdale total about 20 ha, about 0.6% of its original area, but they sample a range of environments and are important in an area that is almost completely built-up[18]. Because of the pressures on them, careful management is needed for their survival.

14 South Sydney

South Sydney covers suburbs from Redfern and Chippendale to St Peters and Roseberry, Surry Hills, Darlinghurst and Woolloomooloo, and Moore Park and the Showground. It is now a densely populated residential and industrial area with virtually no remaining natural vegetation. At the time of European settlement the western part, from Chippendale to St Peters would have had Turpentine-Ironbark Forest on its Wianamatta Shale soils. Clearing of these areas began soon after settlement because of their proximity to Sydney and the reasonably good soils.

On the eastern side of Sydney, at Woolloomooloo, Darlinghurst and Surry Hills, Hawkesbury Sandstone outcropped, its sandy soils being poor agriculturally. Here, in the early nineteenth century, wealthy Sydneysiders built grand houses on large estates, often taking advantage of spectacular harbour views. Sir Thomas Mitchell, Surveyor General of New South Wales, boasted of his Darlinghust property to his brother: 'I have got the most picturesque hill about Sydney, with ten acres of ground around it, for the purpose of building a mansion, which as it will stand on a rock, I am thinking of calling Craigend'[19]. Elizabeth Bay House, built for Alexander Macleay in 1837, was a fine example, though it had a particularly extensive garden, where Macleay grew plants from all over the world amidst remnants of the natural bushland. This bushland would have had trees of Blackbutt, *Eucalyptus pilularis* and Smooth-barked Apple, *Angophora costata*, with shrubby understorey species similar to those still to be found at Vaucluse in Nielsen Park, now part of Sydney Harbour National Park.

Subdivision of the large estates from the 1860s onwards, and their replacement with high density Victorian terraces, destroyed both bushland and established gardens.

The southern part of South Sydney, from Moore Park to Rosebery, was covered with dune sand that stretched across to the coast. Here Eastern Suburbs Banksia Scrub, a low shrubby vegetation with *Banksia aemula* and *Xanthorrhoea resinosa,* grew on the sand dunes, with sedgelands in poorly drained depressions. A plentiful supply of groundwater and cheap land encouraged the spread of industries at Alexandria and Waterloo in the late nineteenth century, expanding further south and east later. To protect the Lachlan and Botany Swamps which provided Sydney's water supply until the 1880s, much of the area remained in public ownership to become available later for public recreation in Moore Park, the Showground and Centennial Park.

Also little used until the late nineteenth century was the tidal Sheas Creek which drained into the Cooks River. With the expansion of industrial activity the upper reaches in Alexandria, previously a muddy swamp with *Juncus kraussii* and *Casuarina glauca,* rapidly became a polluted drain. This was channelled and straightened in the 1890s to become the Alexandra Canal. In 1896 the consultant geologists, Etheridge and David, reported:

View towards Rushcutters Bay in the late nineteenth century. The forest on the alluvium in the valley has been cleared but on the outcropping sandstone here near Kings Cross, woodland of Blackbutt, *Eucalyptus pilularis*, remains. Similar vegetation appears on Darling Point in the distance. (Mitchell Library, State Library of New South Wales)

Previous to the cutting of the present canal and the artificial raising of the level of the surrounding land, the area referred to in this paper [i.e., upstream from the Ricketty Street bridge] was mostly a salt water swamp, through which crept the sluggish malodorous Shea's Creek... The surface of the swamp is covered by rank grass and salsolaceous plants with a thin belt of swamp oak (*Casuarina*) [*glauca*] along its western margin.[20]

An interesting discovery during this work were remnants of an ancient forest, *in situ* tree stumps tentatively identified as *Eucalyptus robusta* buried beneath 4–5 m of tidal-flat mud and indicating that sea level here was once lower.

15 Strathfield

Strathfield Municipality occupies gently undulating country with clay soils from the Wianamatta Shale, and like Burwood and Ashfield further east, the original vegetation was cleared in the nineteenth century for grazing land. At 'Homebush' for example, 'for some hundreds of acres all around not a native tree nor even a stump was visible, so completely had the land been cleared, although not worth cultivation' wrote Louisa Meredith in 1840[21]. The original vegetation would have been Turpentine–Ironbark Forest. Joseph Maiden writing in 1893 indicates that ironbark was abundant: 'I have it on the authority of Mr F H Potts that much of the ironbark used on the Sydney-Parramatta railway came from his father's property between Homebush and Rookwood'[22]. Three species of ironbark were originally found in Strathfield: *Eucalyptus paniculata,* Grey Ironbark; *Eucalyptus siderophloia,* Northern Grey Ironbark; and *Eucalyptus fibrosa,* the Broad-leaved Ironbark. There are still some Turpentines in Strathfield Park and a small patch of mangroves in Mason Park.

Parramatta and the Cumberland Plain

The low rainfall and undulating Wianamatta Shale soils of the Cumberland Plain provide similar natural environments for a group of western Sydney local councils. The area originally supported grassy woodlands and was important from the earliest days of settlement, providing the main grazing and agricultural lands until the western road to Bathurst was opened in 1815. More recently it has received, and will continue to accommodate, the bulk of Sydney's expanding urban population. Because of its particular landscape and history, the condition and problems of remnant native bushland here are somewhat different from those elsewhere in Sydney. In particular there is an urgent need to conserve adequate bushland areas now while they are still available. At present there are virtually no adequate nature reserves on the Cumberland Plain.

16 Parramatta

After crop failures on the poor sandy soils near Sydney, the discovery of better land at Parramatta ensured the colony's survival. The country around Parramatta, or Rose Hill as it was first called, was explored first by Governor Phillip in April 1788. John White, accompanying Phillip, describes the Parramatta River below the present Church Street bridge: 'The banks of it were now pleasant, the trees immensely large, and at a considerable distance from each other; and the land around us flat and rather low, but well covered with the kind of grass just mentioned [i.e., rich and succulent]'[23].

The 'immensely large trees' were probably part of the Cumberland Plain Woodland of Grey Box, *Eucalyptus moluccana,* and Forest Red Gum, *Eucalyptus tereticornis,* with an open grassy understorey that extended westward from Parramatta across the Cumberland Plain. A few remnant trees survive in Parramatta Park. The town itself was sited at the head of navigation of the Parramatta River, taking advantage of the freshwater upstream and boat access below. Here, according to White, 'the tide ceased to flow; and all further progress for boats was stopped by a flat space of large broad stones, over which a fresh-water stream ran'. There would have been mangroves, *Avicennia marina,* below Church Street. Above here the water was brackish or fresh and there would have been Common Reed, *Phragmites australis,* along the stream with paperbarks, *Melaleuca linariifolia,* and Rough-barked Apples, *Angophora floribunda,* on the flats.

South of Parramatta Road, from Granville to Regents Park, the Wianamatta Shale country is more low-lying, the soils having concentrations of

Parramatta was established at the tidal limit of the Parramatta River near the junction of the sandstone and shale. In this watercolour of about 1809, looking westward from the northern shore, sandstone outcrops are evident in the foreground and in the river, where they are being used as a ford. Cumberland Plain Woodland on the shale has been partly cleared for the town and surrounding farms. (Mitchell Library, State Library of New South Wales)

ironstone and being generally poorer. Here the western end of the Turpentine–Ironbark Forest that stretched along Parramatta Road from the inner western suburbs mixed with the Cumberland Plain Woodland. This forest had a greater variety of shrub species than the woodlands on the better agricultural country of Rose Hill, but was less useful to the settlers. Along Duck River, remnants of this vegetation have survived at South Granville, where there are roughly 11 ha of bushland on the west bank of Duck River between Wellington and Everley Roads. The south-western and central parts of the bushland are open-scrub and open-heathland dominated by the paperbarks, *Melaleuca decora, Melaleuca nodosa* and *Melaleuca styphelioides* and tend to be richer in small shrub and herb species than much of the woodland. The bushland in the northern part is more open as the clumps of *Melaleuca* and eucalypts are interspersed with a grassland of Kangaroo Grass, *Themeda australis.* The forest structure has been altered by past land use and the plant communities are representative of the different types of secondary regrowth which, by the mid-nineteenth century, had probably displaced the original forest, here mainly of Broad-leaved Ironbark, *Eucalyptus fibrosa,* Grey Box, *Eucalyptus moluccana,* Woollybutt, *Eucalyptus longifolia,* Red Mahogany, *Eucalyptus resinifera,* Stringybark, *Eucalyptus globoidea* and Rough-barked Apple, *Angophora floribunda.* There is fine stand of Cabbage Gum, *Eucalyptus amplifolia,* around the gully towards Wellington Road, and a most unusual occurrence of Grey Gum, *Eucalyptus punctata,* in the central eastern part of the reserve, associated with an outcropping of the Minchinbury Sandstone.

This Duck River bushland is of particular value because of its rich assemblage of native species, particularly those representative of the flora of this eastern part of the Cumberland Plain. Some 264 native species have been recorded, far more than have been found on other nearby reserves. Particularly worthy of mention are the clumps of the rare *Acacia pubescens,* the stand of Grey Gums, *Eucalyptus punctata,* and the unusually high number of 51 native grasses found here. The bushland, other than within the river channel, is remarkably free of exotic weeds, and deserves to be better protected and managed than it is at present.

North of Parramatta, sandstone country is exposed along the Parramatta River and creeks draining to it, Toongabbie and particularly Darling Mills Creek. A side creek, Hunts Creek, dammed last century for Parramatta's water supply, now

The small paperbark trees, *Melaleuca decora* and *Melaleuca nodosa,* are conspicuous in the Grey Box woodland along the Duck River in Everley Park, South Granville. Despite its small size and history of disturbance, this small bushland remnant contains a surprising variety of species and deserves careful management. (J. Plaza, RBG, 1990)

provides Lake Parramatta, the surrounding bushland of which has excellent examples of the typical Blackbutt, *Eucalyptus pilularis,* forest and woodland of the sandstone soils. There is a healthy and diverse shrub understorey with over 200 native species. Similar vegetation occurred near Hannibal Macarthur's estate, 'The Vineyard', overlooking the Parramatta River at Rydalmere, as the naturalist George Bennett describes in 1834:

> The woods in the vicinity of the 'Vineyard' abounded with numerous plants of the *Orchideae* family, growing in a very barren soil. One of these, that has received the colonial appellation of '*native hyacinth*', was just developing its beautiful caerulean blossoms [probably *Thelymitra*], and another its flowers of a bright yellow, spotted internally with brown [probably *Diuris*]. These latter *Orchideae* are named '*boyams*', having their bulbous roots filled with a viscid mucilage, which renders them an article of food among the aborigines: they are also sought after by the colonial children, who are fond of collecting and eating them; the little creatures would readily recognise their favourite '*boyams*' among the specimens I had collected.[24]

North-east of Rydalmere, Parramatta included areas of Turpentine–Ironbark Forest with Blue Gum High Forest in more favourable situations. High Forest remnants survive at Mobbs Hill, where Blackbutt and Sydney Blue Gum trees grow with a shrubby understorey including *Polyscias sambucifolia, Breynia oblongifolia, Platylobium formosum, Clerodendrum tomentosum* and *Exocarpos cupressiformis*. Some of these areas will be protected within the proposed Carlingford Botanic Parkland.

17 Blacktown

Blacktown lies between Parramatta and South Creek, mainly on gently undulating Wianamatta Shale soils, but with some Tertiary-age alluvial deposits along the western edge and Recent alluvial deposits along the creeks. On the shale soils, the original vegetation was woodland of Grey Box, *Eucalyptus moluccana,* and Forest Red Gum, *Eucalyptus tereticornis,* with some Narrow-leaved Ironbark, *Eucalyptus crebra*, on the hills. The understorey was grassy with patches of shrubs. 'The face of the country where they slept, and for several miles in their road, was a poor soil, but finely formed, and covered with the stately white gumtree,' reported John Hunter of the country to the east of South Creek in 1791[25]. He was probably referring to the presence of *Eucalyptus tereticornis.*

Clearing the trees began soon after this, and by the end of the nineteenth century, Blacktown had a mosaic of small farming, grazing and wooded lands reflecting local soil conditions. Between Parramatta and Seven Hills, the 1879 *Railway Guide of New South Wales* reported:

> The country now becomes more interesting, frequent orange groves of dark green foliage, decked with 'the golden fruit of the gardens of the Hesperides,' imparting a new and delightful charm to the beauty of an everchanging landscape. Well grassed apple-tree [*Angophora floribunda*] flats, with undulating and more open country, and farm-houses, gardens, and cottages succeed; until after a run of 6 miles the train stops for a moment at the quiet little rural station of Seven Hills.[5]

Between here and Blacktown, 'as you come along pretty bits of scenery may here and there be observed; open, partially wooded hills — with occasional signs of cultivation, farms, ponds, orchards, and flats — appear to the right, and orange groves and pretty country residences are unfolded to the left'. And so it remained until the post-World War II spread of fibro bungalows, and later brick-veneer houses, swept this rural countryside away.

The largest area of surviving native vegetation is on Water Board land around Prospect Reservoir. This is mainly woodland on shale soils, of Grey Box and Forest Red Gum, with a localised occurrence of Spotted Gum, *Eucalyptus maculata,* and is a very important remnant of the Cumberland Plain Woodland. An interesting aspect of these Prospect woodlands is the large number of pea family (Fabaceae) species, some of which are now very rare. These and a number of uncommon herbaceous species have survived because of the Water Board's protective management. Nurragingy Recreation Area and Kareela Reserve along Eastern Creek (Doonside) also have good examples of Cumberland Plain Woodland and creek vegetation on shale, and are being actively promoted for passive recreation. Care will be needed to ensure that parking and picnic facilities are not extended at the expense of bushland. Mowing of understorey plants to further develop picnic areas is a major, though often unrecognised, cause of bushland loss.

Other remaining native vegetation in Blacktown is part of the Castlereagh Woodlands and is found on the Tertiary alluvial deposits between Windsor and Penrith. These deposits have quite a different vegetation from that on the Wianamatta Shale soils; in particular the soils are very infertile and the natural vegetation is much shrubbier. Blacktown contains the margins of these deposits, and the vegetation sometimes has an intermediate nature, characterised by an intermixing of species from the Tertiary alluvium and the Wianamatta Shale. Frequently there are trees of Grey Box, *Eucalyptus*

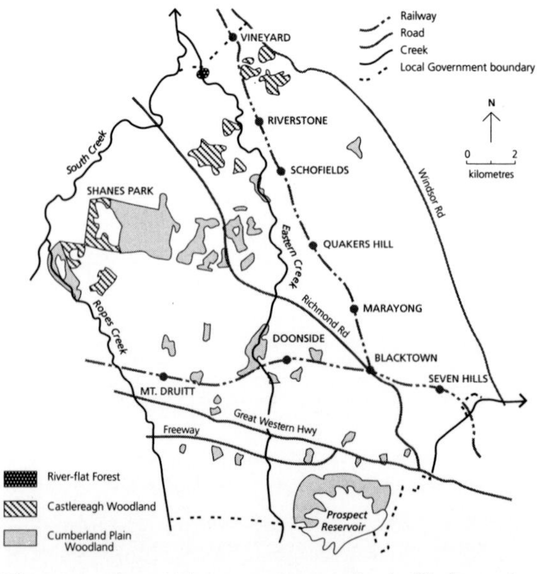

The extent of remaining natural vegetation in Blacktown in 1986. The importance of the vegetation on the Commonwealth land at Shanes Park, one of the largest areas of woodland remaining, can be appreciated.

The characteristic papery bark of *Melaleuca decora*, a common shrub in the Cumberland Plain Woodlands, photographed at Marsden Park. (1975)

moluccana, and Broad-leaved Ironbark, *Eucalyptus fibrosa,* with a shrubby understorey, particularly of paperbarks, the larger *Melaleuca decora* and the smaller *Melaleuca nodosa.* At Marsden Park there is intermediate vegetation but in the western parts of the suburb the vegetation is typical of the Castlereagh Woodlands; Vineyard also has good areas of transitional vegetation including the rare *Grevillea juniperina,* though this is being increasingly disturbed by housing and industrial development. The most important area however is on Commonwealth land at Shanes Park where a very extensive area of bush has survived within the former transmitter station grounds. It should be protected from future development. These areas are rich in native species, and the infertile soil has provided a buffer against exotic weed invasion. Unfortunately elsewhere subdivision for hobby farms and larger suburban blocks is leading to dismemberment of bushland before adequate conservation reserves can be established.

An important remnant of Cumberland Plain vegetation at Shanes Park contains both Cumberland Plain Woodland on shale and Castlereagh Woodland on Tertiary alluvium, shown exposed here on the left. (1974)

Cumberland Plain Woodland at Prospect. Trees are Grey Box, *Eucalyptus moluccana* and the shrubs are Blackthorn, *Bursaria spinosa*. A number of unusual ground cover species occur here. (1975)

18 Camden

Camden Municipality is centred on the town of Camden, on the banks of the Nepean River, in the south-west part of the Cumberland Plain. The district, with its rich fertile Nepean River flats and rolling Wianamatta Shale hills, was named 'The Cowpastures' in the 1790s, after cattle that had escaped from Sydney were discovered running wild here. In 1793 David Collins wrote:

> The country where they [the cattle] were found grazing was remarkably pleasant to the eye: everywhere the foot trod on thick and luxuriant grass; the trees were thinly scattered and free from underwood, except in particular spots; several beautiful flats presented large ponds, covered with ducks and black swan, the margins of which were fringed with shrubs [including *Melaleuca linariifolia Melaleuca styphelioides, Acacia floribunda, Hymenanthera dentata*] of the most delightful tints, and the ground rose from these levels into hills of easy ascent.[26]

'Manangle' and 'Barragal' were names given by the Aborigines to two of the lagoons that would have provided plentiful food for them.

Camden and its reputation is synonymous with that of the Macarthur family. John Macarthur managed to secure 5,000 acres (2,000 ha), described as 'dry, firm and in every Respect so well adapted for a Sheep Pasture'[27] at Mount Taurus. This was to become the great colonial estate of 'Camden Park'. Part of the Cumberland Plain Woodland, the original vegetation, was characterised particularly by trees of Grey Box, *Eucalyptus moluccana,* Narrow-leaved Ironbark, *Eucalyptus crebra,* and Forest Red Gum, *Eucalyptus tereticornis.* The ground cover was grassy with much Kangaroo Grass, *Themeda australis,* and others including *Danthonia racemosa, Danthonia tenuior, Chloris ventricosa, Poa labillardieri, Aristida ramosa, Sporobolus creber,* and *Bothriochloa decipiens.* Patches of shrubs included *Bursaria spinosa, Myoporum montanum, Olearia viscidula* and *Indigofera australis.* Today areas of this woodland with native pasture still remain in the Elizabeth Macarthur Agricultural Research Institute.

Similar grassy woodland was found on the eastern side of the Nepean around Cobbitty, Narellan, and further north. Also near Cobbitty, about 1 km south-east of Cobbitty Trig, was the 'Native Vineyard', a very localised and intriguing patch of dry rainforest, reported by the Parramatta botanist William Woolls in 1867:

> This spot is also remarkable for the occurrence of many plants which do not grow anywhere in the adjacent bush... The 'Vineyard' as it is called [presumably named because of the predominance of vines and lianes] contains only a few acres, and is almost in sight of the Reverend T Hassells's residence at Denbigh. It has the

Fig. 7 This early illustration of *Platylobium formosum*, common in Blue Gum High Forest and Turpentine–Ironbark Forest, is typical of those illustrations used to publicise the first discoveries of the Australian flora.

Fig. 9 River-flat Forest with River Peppermint, *Eucalyptus elata*, near Camden.

Fig. 8 John Lewin's *View of the River Hawkesbury*, probably somewhere near Wilberforce, shows how much of the River-flat Forests has already been cleared by 1805. (Dixson Galleries, State Library of New South Wales)

Fig. 10 Broad-leaved Ironbark, *Eucalyptus fibrosa*, and colourful shrubs in the Castlereagh Woodland at Castlereagh State Forest.

Fig. 11 The Government Hut at The Cowpastures in 1804 is shown surrounded by woodland of dark-trunked eucalypts, most likely Narrow-leaved Ironbarks, *Eucalyptus crebra*. Similar woodland still survives on the Elizabeth Macarthur Agricultural Institute, part of John Macarthur's 'Camden Park' estate. (Mitchell Library, State Library of New South Wales).

Fig. 12 Characteristic Cumberland Plain Woodland trees: Grey Box, *Eucalyptus moluccana* and Forest Red Gum, *Eucalyptus tereticornis*, foreground, together with Narrow-leaved Ironbark, *Eucalyptus crebra*, in the background and an understorey of shrubs and grasses at Mount Annan Botanic Garden.

Fig. 13 Shrubby heath with *Allocasuarina distyla* grows in pockets of shallow soil amongst woodland on this sandstone ridge-top near Bantry Bay.

Fig. 14 This open-forest in Pennant Hills Park is typical of the vegetation on sheltered sandstone slopes in many of Sydney's bushland reserves.

On the Cowpasture Road, 1821–1823, by Edward Mason, showed a typical Cumberland Plain woodland scene. (Mitchell Library, State Library of New South Wales)

appearance of a cultivated shrubbery, rather than that of a collection of plants formed by natural causes.[28]

Some of the plants Woolls recorded have not been found elsewhere near Sydney, the nearest records for several being in the Hunter Valley or Albion Park. Alas, the 'Vineyard' has been degraded, by competition from the naturalised African Olive, *Olea europaea* subspecies *africana*, and grazing by domestic stock and rabbits, and most of the native trees and vines on Woolls' list are now either extinct or moribund old plants.

Along the Nepean River tall eucalypt forest grew on the alluvial banks and freshwater swamps in the flood-prone swales. 'The banks of the Cowpasture river near Narellan are high, sandy, and clothed with goodly gum-trees, swamp oaks and scrubby brushwood,' wrote Peter Cunningham in 1827[13]. Most of this forest disappeared soon after settlement, but at 'Camden Park' immediate river bank forest was fenced off and is now some of the best remaining alluvial forest left along the Nepean River. Here are venerable old Blue Box, *Eucalyptus bauerana,* and Broad-leaved Apple, *Angophora subvelutina*, trees over two hundred years old, with younger Ribbon Gums, *Eucalyptus viminalis* and River Peppermints, *Eucalyptus elata.* Bangalays here are intergrades of *Eucalyptus botryoides* and *Eucalyptus saligna,* some 25 m high.

Grey Box trees, *Eucalyptus moluccana*, line a neat country road somewhere near Camden at the turn of the century. Post-and-rail fencing of ironbark timber, and Bunya Pines, *Araucaria bidwillii*, planted around the homesteads were characteristic features of the Cumberland Plain landscape in the nineteenth century.

Trees of Narrow-leaved Ironbark, *Eucalyptus crebra* (left) and Forest Red Gum, *Eucalyptus tereticornis*, are still common around Camden and Campbelltown though the native shrub and ground cover species are rapidly disappearing. (1975)

River Peppermints, *Eucalyptus elata*, and Bangalay/Blue Gum intergrades, *Eucalyptus botryoides/saligna*, grow to 25 m in this small remnant of River-flat Forest at Camden Park, Menangle. (1988)

There are River Oaks, *Casuarina cunninghamiana,* and Water Gums, *Tristaniopsis laurina,* along the water's edge. The changing flooding patterns and the increasing nutrients in the river have enabled contingents of exotic weeds to establish and naturalise in the understorey, though there were still 97 different native species there recently. The presence of some of the naturalised exotic species results from the activities of John Macarthur's son, Sir William Macarthur (1800–82), a pioneer in horticulture and Australian botany who introduced many new plants to the gardens, arboretum and paddocks at 'Camden Park'. African Olive and other species such as Honey Locust, *Gleditsia triacanthos,* Nettleberry, *Celtis occidentalis,* and Broad-leaved Privet, *Ligustrum lucidum*, were introduced as hedge plants and have subsequently become weeds, naturalised in many places along the Nepean. Sir William also drew the attention of the eminent Victorian botanist Ferdinand von Mueller to the distinctiveness of the now rare Camden White Gum, *Eucalyptus benthamii,* as a separate species, and sent material of it and other

This mature stand of Blue Box, *Eucalyptus bauerana*, and *Angophora subvelutina*, part of the River-flat Forest at Camden Park, will regenerate naturally if grazing of tree seedlings is prevented. (1988)

eucalypts to Paris for the 1861 Exhibition. Camden Woollybutt, *Eucalyptus macarthurii,* another rare species was named after Sir William, though the species grows naturally around Moss Vale, not Camden.

Freshwater swamps filled the low-lying floodplain depressions — Barragal, Menangle and Belgenny Lagoons — but on the low-lying flats around Cawdor and Narellan was Swamp Oak, *Casuarina glauca*, forest. This is usually a species of coastal estuaries, often found near mangroves, but at Camden it appears to grow in response to saline ground water. Major Thomas Mitchell wrote in 1839 that after crossing the Nepean River and entering the County of Cumberland, 'the soil is good and appears well cultivated, but there is a saltness in the surface water, which renders it, at some seasons, unfit for use'[29].

Camden municipality is still largely rural, but with the inevitable growth of the suburban populations nearby, remnants of bushland will be increasingly threatened. Some of the best areas are within the old 'Camden Park' estate and it should be the responsibility of the N.S.W. Department of Agriculture and Fisheries to ensure that these areas are protected and maintained.

An old Narrow-leaved Ironbark, *Eucalyptus crebra*, near Campbelltown. Young saplings have been able to establish and grow following the cessation of grazing. (1975)

19 Campbelltown

The town of Campbelltown on the south-eastern edge of the Cumberland Plain was established by Governor Macquarie in 1820. To the east lay rugged Hawkesbury Sandstone country; to the north and west, the rolling hills with woodlands of Grey Box, *Eucalyptus moluccana,* Forest Red Gum, *Eucalyptus tereticornis,* and Narrow-leaved Ironbark, *Eucalyptus crebra,* on the clay soils of the Wianamatta Shale. The naturally grassy ground cover was soon being cleared to grow wheat and other cereals, and for the next 40 years Campbelltown was part of the 'granary of the colony'. However, in the 1860s wheat stem rust fungus destroyed the wheat industry here. Dairying became the major agricultural industry, together with fruit-growing along the Georges River at Minto and Wedderburn.

Nearly 200 years of agriculture and the recent mushrooming of Campbelltown as a major urban centre have left little of the original shale woodland. A Grey Box-Forest Red Gum woodland remnant has been retained in the new Mount Annan Botanic Garden, where despite a long history of dairying, several now rare local native species have been found. Research into floristic changes following the cessation of grazing, the effects of fire and the invasion of exotic weed species will be carried out here. Beside the

Spotted Gum, *Eucalyptus maculata*, occurs naturally in dry woodland at Appin, Hoxton Park and The Oaks. Another population is found in wetter forest on Narrabeen Group shales on the Palm Beach peninsula. (1975)

Woodland of Grey Box, *Eucalyptus moluccana*, and Forest Red Gum, *Eucalyptus tereticornis*, in the Woodland Conservation Area at Mount Annan Botanic Garden at Campbelltown. Natural regeneration of native species is being encouraged. (1986)

Appin Road, just north of Appin is an interesting stand of the readily recognisable Spotted Gum, *Eucalyptus maculata,* with a shrubby understorey of *Acacia*. Isolated from its other nearest natural occurrences at Hoxton Park and Werombi, this naturally occurring population should be preserved. The expansion of Campbelltown as an urban growth centre is now destroying the last remnants of the shale woodland flora.

In contrast to the shale country, the poorer sandstone lands along the Georges River east of Campbelltown, and the river gorge itself, were not used for agriculture. Now, however, the growing suburbs of Campbelltown are pushing out rapidly; Minto Heights, Kentlyn, Airds, Wedderburn. These new suburbs typically occupy the plateaus and ridges, leaving the creek lines and river gorge for recreation and conservation. Inevitably stormwater run-off from suburban areas is directed into these gullies, promoting weed invasion, particularly by Privet, *Ligustrum sinense,* and leading to the deterioration of the native low-nutrient-favouring sandstone species. Different stages in the sequence, from slightly weed-infested to totally weed-infested, can be recognised throughout Sydney and can generally be related to the increasing age of the adjacent suburbs. The only really successful way small conservation reserves can be protected in the long-term is by inclusion of the complete local subcatchment within the reserve.

East of the Georges River, Hawkesbury Sandstone woodland and forest extends for miles over the Holsworthy Firing Range and the Woronora Catchment. Protected from human intrusion, apart from exploding missiles and four-wheel-drive army vehicles, bushland here has remained otherwise undisturbed. It is an area rich in plant species but has never really been explored botanically. What exciting plants await discovery here?

20 Fairfield

Fairfield City Council, situated between Prospect Reservoir and the Georges River, is confined to the undulating country of the Wianamatta Shale. On the western side, around Cecil Park and Horsley Park, the country is quite hilly and the original vegetation was a forest of Spotted Gum, *Eucalyptus maculata,* and Grey Box, *Eucalyptus moluccana.* 'Timber growing of valuable qualities, one of which (Spotted Gum) is now daily rising in repute for Staves of Casks,' ran an advertisement in the *Sydney [Morning] Herald* in November 1831 for a farm watered by Orphan School Creek, probably near Edensor or Cecil Park[30]. As most of this forest has now been cleared, it is important that remnants here, particularly along Elizabeth Drive, are protected. The Bossley Bush Recreation Reserve, for example, has a small but fairly intact example of this forest. Other isolated populations of Spotted Gum on the Cumberland Plain are at Appin and Werombi; it also occurs in quite different moist tall forest along the northern Warringah Peninsula.

Elsewhere, on the undulating to level country on the western side of Prospect Creek, Fairfield's vegetation was typical Grey Box-Forest Red Gum Cumberland Plain Woodland, together with Cabbage Gum, *Eucalyptus amplifolia,* paperbarks, *Melaleuca decora,* and Swamp Oak, *Casuarina glauca,* in low-lying poorly drained sites. Early colonial settlement along the Georges River spilled into Fairfield. For example, in 1824 Joseph Lycett reported, 'some of the very best land in this part of the Country is situated at *Cabramatta*'[8]. Woodlands further from the early settlement sites were cleared for market gardens and poultry farms in the late nineteenth century, though there are still remnants along watercourses such as Prospect Creek.

Suburbs and industry spread through much of Fairfield after World War II.

On the eastern side of Prospect Creek, towards Bankstown, and now almost completely built on, was 'tea tree scrub' with *Melaleuca decora* and *Melaleuca nodosa* and a greater variety of shrub species than on the normal clay soils. Much of this scrub remained essentially undisturbed until its post-World War II transformation into the industrial complexes of Old Guildford and Yennora, and the residential suburbs of Villawood and Carramar.

21 Hawkesbury

The fertile alluvial flats along the Hawkesbury River at Windsor were discovered in 1791. 'The banks are high, and the soil a light sand, but producing fine straight timber,' wrote John Hunter at the time[25]. The 'fine straight timber' would probably have been trees of *Eucalyptus tereticornis,* Forest Red Gum, but also called Flooded Gum by the early colonists. This appears to have been the predominant tree in the River-flat Forest of the Hawkesbury, particularly between Richmond and Windsor. There may also have been some *Eucalyptus deanei,* Deane's Gum, or *Eucalyptus saligna,* Sydney Blue Gum, though there is no direct evidence for these species on the main floodplain. They are now found along the Grose and Colo Rivers respectively.

Beneath the forest trees would have been grasses, such as *Microlaena stipoides, Stipa verticillata* and *Stipa ramosissima,* shrubs such as *Bursaria spinosa* and *Hymenanthera dentata,* ferns such as *Pteridium esculentum,* and vines such as *Smilax* and *Eustrephus.* James Atkinson writes in 1826: 'In alluvial lands, a kind called blady grass is found; this is a very coarse variety the ribband being half an inch wide, and it is probably not very nutritive'[31]; this sounds very like the Blady Grass, *Imperata cylindrica,* now common in frequently burned places. Atkinson's observations on the abundance of Blady Grass, made about 30 years after first settlement of the floodplains, may indicate a response to the settlers' use of fire to clear and maintain these sites, or may indicate that frequent burning of the floodplains had been carried out by the Aboriginal peoples, or perhaps both.

Associated with the floodplain are the typical back-swamps and lagoons of the Hawkesbury. Most of these have been altered, either drained and only holding water for short periods after flooding or heavy rain, or dammed to become permanent swamps. Longneck Lagoon at Pitt Town, for example, had permanent water with some emergent reeds and *Eleocharis sphacelata* with *Juncus usitatus* rushland on the periodically wet margins in the 1970s, but a permanently higher water level and apparent changes in water quality in the 1980s are

Longneck Lagoon *(left)* in 1972, and *right)* in 1989. Waterplant diversity has deteriorated noticeably during this period, apparently related to a higher, non-fluctuating water level and decreased water quality.

leading to the loss of this vegetation and deterioration of the lagoon. As it is the site of a field studies centre and important conservation area, efforts need to be made to rectify the changes. Pitt Town Lagoon, though a nature reserve and an important site for waterbirds, has been drained, and its native vegetation replaced with pasture grass. Other swamps are found further downstream in Baulkham Hills shire.

Ironbark forest with a variety of shrub species is an important part of the natural vegetation in the catchment of Longneck Lagoon. (1972)

The five Hawkesbury towns of Windsor, Richmond, Wilberforce, Pitt Town and Castlereagh were established in 1810 by Governor Macquarie, to promote the agricultural development of the floodplain. By 1826 James Atkinson could write: 'The greater part of the alluvial lands upon the Hawkesbury and Nepean have been cleared, and are under cultivation'[31]. Old photos and paintings of the Hawkesbury during the nineteenth century show an agricultural landscape similar to that of today.

Back from the floodplain the higher country is on Wianamatta Shale, Tertiary alluvium or Hawkesbury Sandstone. The Wianamatta Shale areas between Oakville and Cattai had typical grassy woodlands of the Cumberland Plain, with Grey Box, *Eucalyptus moluccana,* Forest Red Gum, *Eucalyptus tereticornis* and Narrow-leaved Ironbark, *Eucalyptus crebra.* The best remaining example of this woodland is at Scheyville near Pitt Town, within the catchment of Longneck Lagoon. Here the woodland has a healthy diverse understorey including rare species and a varied creekside flora; as one of the most important examples of Cumberland Plain vegetation, it was proposed as a nature reserve in 1975. However, there has been conflict over its future because it is Crown land and is seen as a cheap site for housing. When there is so much adjacent cleared freehold land of equal suitablility, the destruction of significant bushland can only be regarded as a short-sighted action that will diminish the quality of the future suburban environment.

At South Windsor on the Tertiary alluvial deposits is woodland of Grey Box and Broad-leaved Ironbark, *Eucalyptus fibrosa,* while an outlier at Pitt Town has woodland of Scribbly Gum, *Eucalyptus sclerophylla,* and Broad-leaved Ironbark with a rich shrub understorey. These are part of the Castlereagh Woodlands, best seen around Castlereagh State Forest and Agnes Banks further south. Part of the Riverstone Meat Company's land between South Windsor and Riverstone has been proposed for a nature reserve, but at time of writing definitive action has not been taken.

Hawkesbury Shire also includes very extensive wooded sandstone country north and west of the Hawkesbury River, outside the scope of this book.

22 Holroyd

Holroyd is a small Cumberland Plain municipality south-east of Prospect Reservoir, mostly on undulating Wianamatta Shale country with deep, often poorly drained clay soils. Prospect Hill, a prominent feature formed from an igneous dolerite intrusion of Jurassic age, is the exception, and a landmark for the first European exploring parties. 'We came to a pleasant hill, the top of which was tolerably clear of trees and perfectly free from underwood,' wrote Surgeon-General John White, accompanying Governor Phillip in April 1788. 'His excellency gave it the name of Belle Veue'[23]. From the same hill, a year later, another exploring party with Captain Watkin Tench reported 'Before us lay the trackless immeasurable desert, in awful silence'. Heading north-west the party 'continued to march all day through a country untrodden before by any European foot. Save that a melancholy crow now and then flew croaking over head, or a kanguroo was seen

Old Prospect Road, Greystanes, in 1934. As in neighbouring Blacktown and Fairfield the Grey Box, *Eucalyptus moluccana*, woodlands have been cleared for farms, but there is no sign of the suburban expansion that will follow after World War II. (Government Printing Office Collection, State Library of New South Wales)

The construction of the water pipeline to Prospect in 1896 cut through surprisingly tall forest here at Merrylands West. (Macleay Museum)

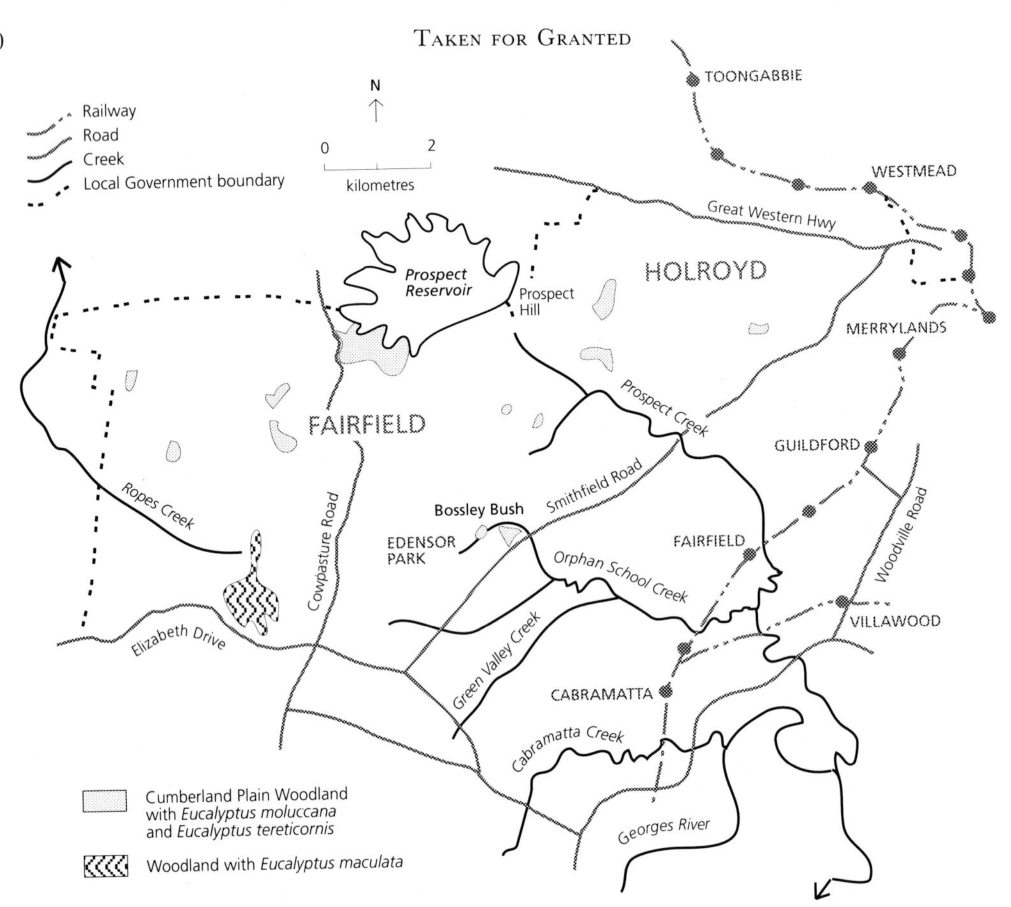

Remaining native vegetation in Holroyd Municipality (north of Prospect Creek) and Fairfield in 1986 includes Cumberland Plain Woodland with *Eucalyptus moluccana* and *Eucalyptus tereticornis*, and woodland of *Eucalyptus maculata*.

to bound at a distance, the picture of solitude was complete and undisturbed'[32].

Surviving woodland on nearby Water Board land (see Blacktown) is on shale, with typical Cumberland Plain species, Grey Box, *Eucalyptus moluccana,* and Narrow-leaved Ironbark, *Eucalyptus crebra,* and a grassy cover with patches of *Bursaria spinosa* scrub. Woodland like this would have continued westward (the 'trackless immeasurable desert') to the river flats of the Nepean-Hawkesbury River.

On the rich volcanic soil on Prospect Hill we would expect there to have been different species from those on the surrounding shale soils. John White indicates that the vegetation was more open, but no records have survived, and the hill has now been largely cleared and quarried for blue metal aggregate. A number of herbaceous species in the woodland around Prospect Reservoir, including *Scutellaria humilis* and species of *Ranunculus,* may be indicative of high nutrient or high pH soils, and suggest that the nearby intrusion of igneous rock may be influencing the surrounding soils.

Most of Holroyd's land has been farmed or grazed since the very early days of settlement, though since World War II it has become increasingly suburban. Apart from the woodland around Prospect and some remnants along Prospect Creek and near Alpha Park, Greystanes, there is very little of the native vegetation now to see.

23 Liverpool

Liverpool City Council area stretches from the Georges River at East Hills to the Nepean River at Wallacia. With its gently undulating Wianamatta Shale country and remnants of the original open-forest and woodland, it provides a cross-section of the Cumberland Plain landscape.

In 1832 however, the naturalist George Bennett wrote as he travelled from Parramatta to Liverpool:

The uncleared land has a dismal appearance; the huge blue gum, stringy bark, box, and ironbark trees, (all of the *Eucalyptus* genus,) rose from the thick bush which surrounded their bases, to a great elevation... The road was in excellent condition, but the land on each side was for the most part uncleared; and, being covered by dense forest trees, had a very sombre character. A few trees of the 'green wattle', (*Acacia decurrens*), profusely covered by golden blossoms, and occasionally a cleared verdant space, alone gave anything like animation to the scenery.[24]

The main trees were Grey Box, *Eucalyptus moluccana,* and Forest Red Gum, *Eucalyptus tereticornis.* On the hilly country around Hoxton Park and Cecil Park the distinctive Spotted Gum, *Eucalyptus maculata,* accompanied these species. On the deeper alluvial soils and poorly drained creek flats, such as along South Creek, Kemps Creek and Orphan School Creek, was a taller forest of Cabbage Gum, *Eucalyptus amplifolia,* and Rough-barked Apple, *Angophora floribunda,* sometimes with patches of Swamp Oak, *Casuarina glauca. Casuarina glauca* is commonly found near estuarine swamps and lakes such as Towra Point or Narrabeen Lagoon, and its discovery away from the coast surprised botanists. Joseph Maiden, director of Sydney's Botanic Gardens from 1896 to 1924 recalled:

A young man was driving me a number of years ago near Edensor Park, Liverpool, and I saw some She-oaks in the distance which looked to me like the Salt-water Swamp Oak (*Casuarina glauca*). (At that time the species was not known so far from sea or a salt-water creek.) I asked the driver what name he gave to those oaks, and he said, 'Oaks'. A little later I was able to collect specimens, and then knew that the trees were truly *C. glauca*. I told the driver that I had never before seen this oak except on the banks of brackish creeks or lagoons, and he said nothing. But he was evidently pondering the matter, for in a few minutes he said that all round about those oaks the water was so salt in a dry time that the cattle could not drink it. Here was the solution. Although far away from the sea — that infinite source of brackishness — there was a stock of salts in the soil hereabout.[33]

Brackish subsoil water is characteristic of the Wianamatta Shale.

Patches of paperbarks, *Melaleuca decora* and occasionally *Melaleuca styphelioides,* also indicate poorly drained sites. A low-lying area of forest at Kemps Creek has been recommended for preservation as a nature reserve to retain an example of the tree and understorey species characteristic of such sites. Elsewhere such vegetation, and particularly the understorey, is being degraded by competition from exotic species such as *Paspalum* and herbs such as Fennel and Dock, by regular mowing and spraying, which again favour exotic grasses, and by channelisation of creeks to form stormwater drains.

Liverpool still has a considerable number of bushland areas but very few are protected; with increasing development, each area will be carved up as expendable unless action is taken soon to protect significant areas.

Near Kemps Creek village, but not in the proposed reserve, are outlying remnants of the Tertiary alluvial gravels similar to those found at Castlereagh. These soils, found as remnants on the broad dividing ridge between Kemps and South

Old trees of the rare Camden White Gum, *Eucalyptus benthamii*, at Bents Basin in 1961 during a film shoot. Fire in 1979 caused severe deterioration.

Creek, have the distinctive Castlereagh Woodland vegetation of Broad-leaved Ironbark, *Eucalyptus fibrosa,* and Scribbly Gum, *Eucalyptus sclerophylla,* and a characteristic shrubby understorey with a range of Myrtaceous and Fabaceous species. A small patch of similar vegetation near the showground at Prestons is also important and should be protected. The soils at these sites are infertile and small remnants of these shrubs can be preserved without the problems of weed invasion common on richer creekside sites.

A remnant of Spotted Gum, *Eucalyptus maculata,* open-forest west of Hoxton Park Aerodrome is an important survivor of the forest once occurring on the steeper shale country here. With its natural understorey of grasses and herbs it provides considerable wildlife value in an area that will become intensively urban.

On its eastern side Liverpool takes in the great bend in the Georges River, including Chipping Norton, Moorebank, Hammondville and Holsworthy. Botanically this is a very varied area but unfortunately one that has been fragmented by development. Along the Georges River are estuarine and freshwater swamp areas. Between East Hills and Newbridge Road are patches of paperbark, *Melaleuca ericifolia,* and Swamp Oak, *Casuarina glauca,* separated from the river by a narrow zone of Grey Mangroves, *Avicennia marina.* The mangroves occupy the intertidal zone, with the Swamp Oak on the occasionally inundated zone and paperbark in areas principally flooded by freshwaters. At Voyager Point opposite East Hills is a rich complex of plant communities including the most undisturbed and largest freshwater swamp on the Georges River, as well as communities on shale, sandstone and Tertiary alluvium. At Chipping Norton gravel and sand extraction has created a number of lakes which, though having only a few plant species, provide good habitat for waterbirds.

At Moorebank and Holsworthy reasonably extensive deposits of Tertiary gravels and sands overly the sandstone. These areas have woodland and low woodland similar to the Castlereagh Woodlands with Scribbly Gum, *Eucalyptus sclerophylla,* and Narrow-leaved Apple, *Angophora bakeri,* and a rich understorey of shrubby species. There are also patches of wet heath with *Banksia oblongifolia* and *Xanthorrhoea minor.*

At the other end of Liverpool is the Nepean River, associated with rich Recent alluvial flats, now almost entirely cleared of vegetation. The original forest here had Cabbage Gum, *Eucalyptus amplifolia,* Broad-leaved Apple, *Angophora subvelutina,* and River Peppermint, *Eucalyptus elata,* together with the rare

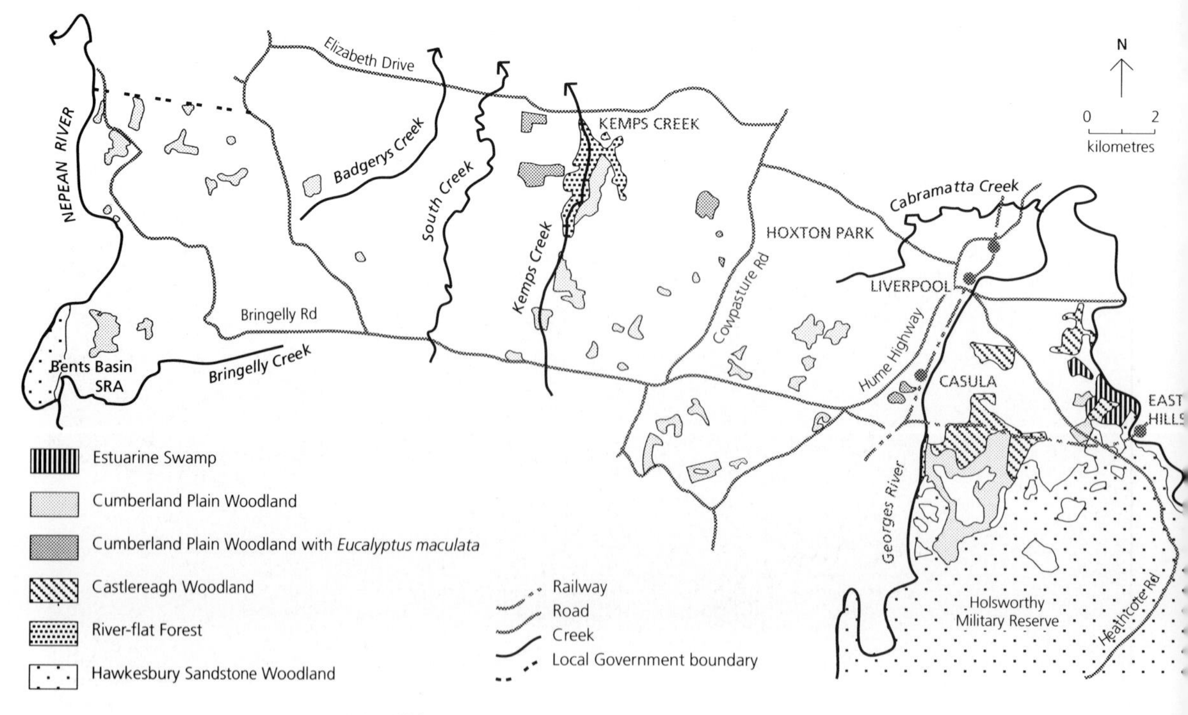

Remaining vegetation in Liverpool in 1986.

Grey Box-Forest Red Gum woodland regenerating vigorously after fire in an area proposed for a nature reserve at Kemps Creek. (1980)

Camden White Gum, *Eucalyptus benthamii*. This last species was once found along the Nepean between Camden and the Grose junction but healthy populations are now found only at Bents Basin State Recreation Area, south of Wallacia, and on Kedumba Creek near Lake Burragorang. As the latter site is threatened by modifications to Warragamba Dam that will raise the level of the lake, careful management of the Bents Basin population is important if the species is to be maintained here in its natural habitat.

Also within Bents Basin State Recreation Area is the spectacular sandstone gorge of the Nepean River from Bents Basin upstream to Campbells Ford. Vegetation along here is quite different from that on the alluvial flats. The sandstone boulders and outcrops have spindly shrubs of *Tristaniopsis laurina,* Water Gum, and *Leptospermum flavescens,* which can bend in the brief but torrential floods that characterise the river. On the banks are trees of Deane's Gum, *Eucalyptus deanei,* and Blackbutt, *Eucalyptus pilularis.* The sandstone 'island', Little Mountain, between the Nepean River and the farming country to the east has a varied woodland and forest vegetation. Tree species include Blackbutt; Yellow Bloodwood, *Eucalyptus eximia;* Grey Gum, *Eucalyptus punctata;* Stringybark, *Eucalyptus eugenioides;* Smooth-barked Apple, *Angophora costata;* Turpentine, *Syncarpia glomulifera;* and Forest Oak, *Allocasuarina*

Remnants of Spotted Gum, *Eucalyptus maculata*, and Grey Box, *Eucalyptus moluccana*, woodland on the road verge at Hoxton Park. Native understorey species still survive on the roadside but have been completely eradicated from the adjacent pasture. (1979)

torulosa. There is a small shale residual with a grassy ground cover and *Eucalyptus crebra,* the Narrow-leaved Ironbark. Bents Basin State Recreation Area is a small (319 ha), but very diverse, reserve with at least 330 native plant species. Because of its size and the proximity of increasing suburban populations, care will be needed to ensure that visitor and recreation pressures do not destroy its important natural values.

24 Penrith

Watkin Tench's description of the Nepean River near Penrith in 1789 includes the first indications of the floods that were to dominate the lives of the colonists along the Nepean-Hawkesbury River:

At daylight we renewed our peregrination; and in an hour after we found ourselves on the banks of a river, nearly as broad as the Thames at Putney, and apparently of great depth, the current running very slowly in a northerly direction... We proceeded upwards, by a slow pace, through reeds, thickets, and a thousand other obstacles, which impeded our progress, over coarse sandy ground, which had been recently inundated, though full forty feet [12 m] above the present level of the river.[32]

Tench also provides evidence that the riverbank vegetation was much denser than on the grassy woodlands of the nearby Cumberland Plain.

River Oak, *Casuarina cunninghamiana,* would have been important here as well as some hardy rainforest species, including White Cedar, *Melia azedarach.* 'The wood of some of these trees is very light; they are about the size of large walnut trees, which they resemble; they shed their leaves, and bear a small fruit which is said to be very wholesome,' reported Governor Phillip, describing trees on the Nepean near Penrith in 1790[34], and almost certainly referring to *Melia azedarach.* It is still common along the river, and has been planted widely in the district. Tench's 'reeds' probably included *Phragmites australis* which was abundant along the Nepean River near Penrith in 1860, and is still common there.

Upstream of Penrith, in the Nepean Gorge that the river has cut through the Hawkesbury Sandstone, the river-flat vegetation is reduced to a fringe of River Oaks, *Casuarina cunninghamiana*, with Deane's Gum, *Eucalyptus deanei* or Blackbutt, *Eucalyptus pilularis,* on the lower hillslopes. On a boat trip to the Nepean Gorge in 1860, George Bennett reports seeing 'some Red Cedar-trees (*Cedrela Australis*) [*Toona australis*], now becoming very rare to the colony; the largest was about 16 feet [5 m] in height'[35]. How widespread Red Cedar was originally on the floodplain is unknown, though it was probably restricted mainly to the banks of the Hawkesbury below Richmond. Bennett's trees were obviously young saplings, and there are still a couple of trees near the junction with Euroka Creek. No other naturally occurring Red Cedar trees appear to remain on the floodplain now.

Grove of Swamp Oaks, *Casuarina glauca*, near Badgerys Creek. Such groves result from the root-suckering ability of this species, and its resilience in changing conditions. (1975)

Penrith, on the eastern bank of the Nepean River where the Great Western Highway crossed, developed as an agricultural town. The fertile river flats near the town were eagerly sought for cultivation, and the tall forests of Broad-leaved Apple, *Angophora subvelutina,* Forest Red Gum, *Eucalyptus tereticornis,* and Cabbage Gum, *Eucalyptus amplifolia,* were soon cleared.

Between Penrith and Parramatta typical Cumberland Plain Woodland of Grey Box, *Eucalyptus moluccana,* Forest Red Gum, *Eucalyptus tereticornis,* and Narrow-leaved Ironbark, *Eucalyptus crebra,* covered the Wianamatta Shale soils. In 1824 the French naturalist René Lesson wrote:

The trees that border the road [from Parramatta] and which compose the entire vegetation of this plain, are Eucalypts, Mimosas, dreary Casuarinas, whose slender, sombre foliage gives a hazy appearance to the forests of which they form part. Certain parts of the road are laid with dolerite taken from an eminence five miles from Parramatta [Prospect Hill]. There is remarkable profusion of growth of Asclepias australis closely related to the Syriaca, and of Mimosa decurrens [*Acacia decurrens*].[36]

The abundance of the introduced weed *Asclepias,* now *Gomphocarpus fruticosus,* was commented on by a number of writers. Mrs Meredith wrote: 'Great quantities of a tall, handsome, herbaceous plant commonly called the "mock-cotton tree" grew near us, and by the roadsides around Sydney'[21]. It is no longer common, possibly due to changes in agricultural practices.

During the nineteenth century the Cumberland Plain Woodlands were gradually cleared to increase grazing lands. An interesting example of selective clearing was Edward Cox's landscape treatment of the entrance drive to his mansion 'Fernhill' at Mulgoa. Here he had all but the locally abundant Rough-barked Apples, *Angophora floribunda,* removed. These trees, with their spreading habit, thick dense leafy crowns and English Oak-like appearance gave his estate a desired park-like landscape[37]. The shape and colour of these trees also led to the name of 'Apple'. North of Mulgoa an interesting remnant of native vegetation close to the shale-sandstone boundary has been proposed as a nature reserve to conserve its interesting shale cliffline habitats, displaying a sequence of geological strata.

Poorly drained sites probably remained uncleared for longest. Stands of *Casuarina glauca,* Swamp Oaks, at Jamisontown indicate poor drainage and the influence of the saline groundwater of the Wianamatta Shale. Swamp Oaks also have the unusual capacity to sucker from roots, and stems can form small groves 10 or 15 m across.

Some patches of eucalypt woodland have also regrown. On military lands at Kingswood, and on the Bringelly Road south of Penrith, stands of trees have regenerated on former grazing lands after the removal of stock. Similarly, the vigorous regeneration of a thicket of seedlings or saplings below old trees of Grey Box, or particularly Forest Red Gum, can be seen on unstocked land such as that awaiting subdivision. Such regeneration is

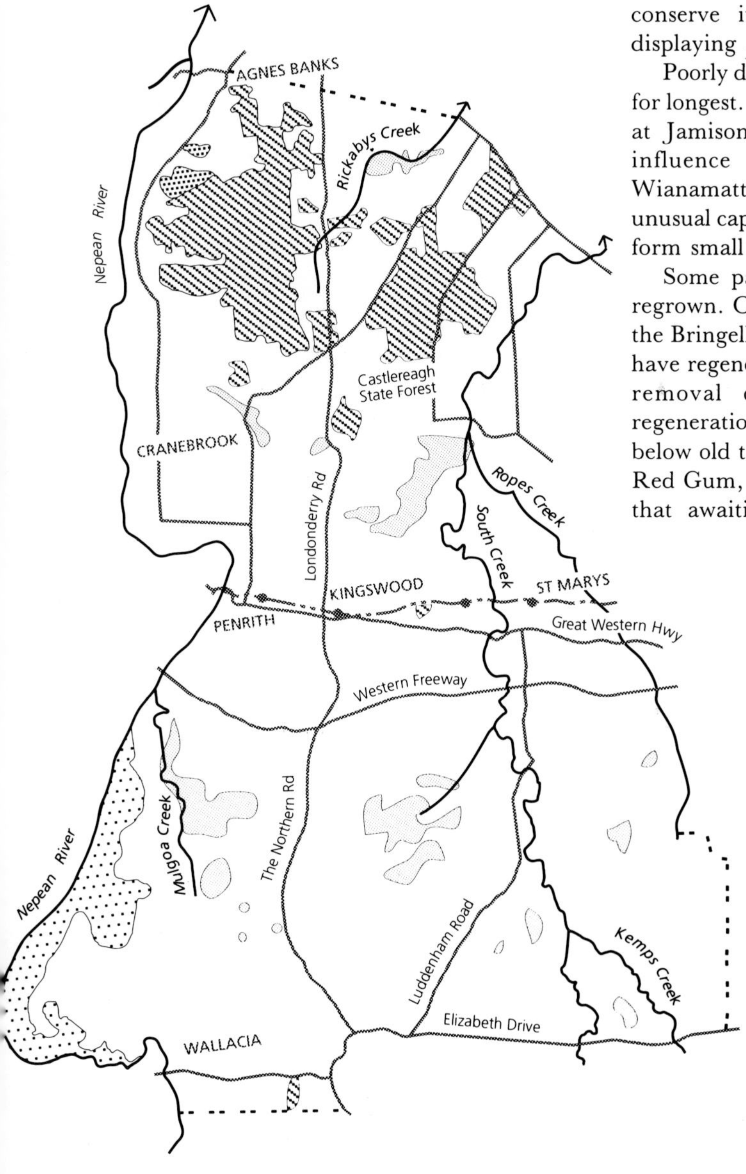

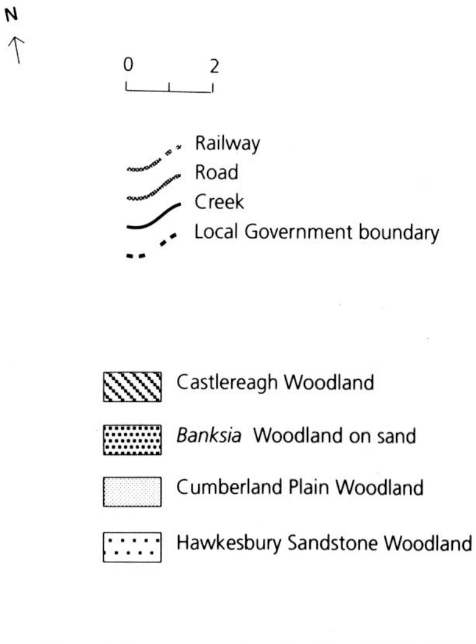

Remaining vegetation in Penrith in 1986.

'Smiths Paddock, Penrith' in 1887. The small trees and grassy, evidently grazed, understorey look similar to many of today's remnants and suggest that prior to 1887 tree-felling and subsequent regrowth had already occurred here. (Government Printing Office Collection, State Library of New South Wales)

Narrow-leaved Ironbark, *Eucalyptus crebra*, is a characteristic Ironbark of the Cumberland Plain Woodlands. This remnant tree in a cleared paddock at Mulgoa at the turn of the century indicates how big some of the original Cumberland Plain trees were.

Today's *Eucalyptus crebra* trees are generally more the size of this one growing with the rare sharp-leaved shrub *Grevillea juniperina* near St Marys. (1979)

Tussocks of *Juncus usitatus* form a rushland on the Nepean River flats near Cranebrook in 1975. The area will be destroyed by the Penrith Lake Scheme. In the background is Grey Box, *Eucalyptus moluccana*, woodland on the shale hillside.

Eucalyptus fibrosa, Broad-leaved Ironbark, shown here in the Castlereagh State Forest, is an important part of the Castlereagh Woodlands. (1977)

Eucalyptus parramattensis is characteristic of poorly drained sites in the Castlereagh Woodlands. (1974)

Woodland of *Eucalyptus sclerophylla*, Scribbly Gum, and *Angophora bakeri*, Narrow-leaved Apple, with a mixed understorey of grasses and low shrubs, is part of the Castlereagh Woodlands. It grows where soil is more sandy. (1977)

almost always within 30 m of the adult tree as eucalypt seed is poorly dispersed and only blows a short distance.

North of Penrith, between Castlereagh and Richmond, are extensive sand and clay deposits of Tertiary age. These have distinctive Castlereagh Woodlands. The species composition depends on the soil, with Ironbarks characterising the clays and Scribbly Gums the more sandy soils. There are two very important natural areas here. The Agnes Banks Nature Reserve preserves a small remnant of the characteristic *Banksia* woodland that once covered an isolated area of white sand dunes. The vegetation is quite different from the surrounding Grey Box and Ironbark woodlands, and is more similar to coastal dune vegetation. Apart fron the Nature Reserve, and a small adjoining area now protected by a Permanent Conservation Order, the vegetation has been destroyed by sand extraction. To the south of this reserve there are important areas of swampy open woodland with *Eucalyptus parramattensis* and wet heathy sites, fairly rich in species. It is important that part of this area is added to Agnes Banks Nature Reserve. Nearby vegetation is threatened with destruction for a new waste dump.

The other important natural area is the Castlereagh State Forest, now a Demonstration Forest [38], and containing the best remaining areas of Castlereagh Woodlands. Here are stands of Broad-leaved Ironbark, *Eucalyptus fibrosa,* with an understorey of shrubs of *Acacia elongata, Dillwynia tenuifolia, Pultenaea parviflora, Micromyrtus minutiflora* and *Dodonaea falcata.* Lower more open woodland of Scribbly Gum, *Eucalyptus sclerophylla,* has shrubs of *Banksia spinulosa, Pimelea linifolia* and *Grevillea mucronulata.* Some of the species are only found in this area. Because the soils here are very poor, the vegetation remained virtually undisturbed, apart from local timber-cutting and gravel extraction, until pressures for hobby farm and suburban housing began in the 1960s.

With the increasing land-use pressures, adequate conservation of this important vegetation is far from assured. Since World War II, Penrith has become a centre for major suburban development, and housing estates have rapidly replaced its rural lands. Native vegetation that survived on the farms or on areas of poor soil is disappearing in the subdivisions but, at the same time, the increasing residential populations are alienated from the natural world. Bushland for recreation and education will be a prime need in the future, but unless we protect adequate areas now, it will not be there when it is needed.

Scribbly Gum, *Eucalyptus sclerophylla,* and *Banksia aemula* on deep sand at Agnes Banks. Good stands of this low woodland have been protected in the Agnes Banks Nature Reserve and adjacent area covered by a Permanent Conservation Order. (1975)

Sedgeland on deep sand at Agnes Banks in 1975. It has now been completely destroyed by sand extraction.

The Eastern Suburbs

The Eastern Suburbs lie south and east of the city, between Port Jackson and Botany Bay. Here are some of the most prestigious suburban residences in Sydney, as well as high and low density residential development, and industrial, recreational and sporting areas. However, despite the high levels of development, there are still remnants of the natural environments — some securely preserved, others with a more precarious future. Yet enough clues remain to piece together a picture of the original plant communities and how they have been instrumental in developing the patterns for this suburban region.

25 Botany

Botany is an industrial and residential municipality on the northern side of Botany Bay between Kingsford Smith Airport and Bunnerong Road. The main features of its original environment were extensive sand dunes draining into a system of freshwater swamps. On these wind-blown sand dunes, which extended from behind the coastal beaches westward to the estuarine flats of Sheas Creek and Cooks River, was the most characteristic vegetation of the Eastern Suburbs, at least up until early this century. Here, covering suburbs like Rosebery and Botany was the Eastern Suburbs Banksia Scrub — varied heath, scrub and low forest

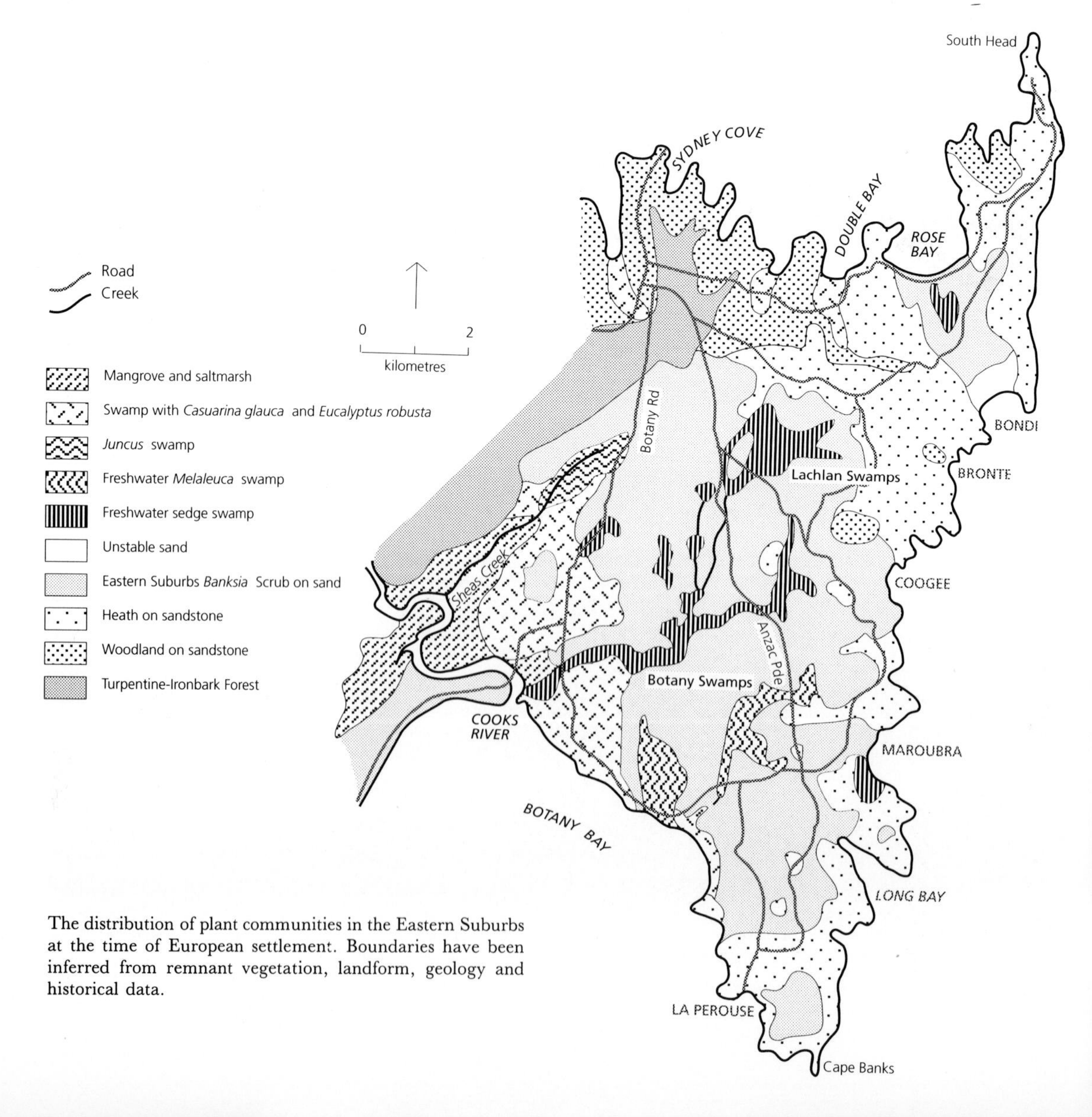

The distribution of plant communities in the Eastern Suburbs at the time of European settlement. Boundaries have been inferred from remnant vegetation, landform, geology and historical data.

vegetation with a rich variety of shrubs, including *Banksia aemula, Monotoca elliptica, Eriostemon australasius, Ricinocarpos pinifolius* and *Xanthorrhoea resinosa.*

'Walked yesterday to Botany Bay, a distance of 8 or 10 miles from our lodgings,' recorded Frederick Mackie in the 1850s:

The road lies over low sand hills covered with small scrub and various flowers. The sand in many places has almost the whiteness of snow and so little mixture of earth is there in it that it would doubtless be entirely destitute of vegetation but for the moisture of it; water is found about 2 ft. below the flat surface. The moister places were generally pink with the flowers of Sprengelia Incarnata, intermixed with Boronias, Bauera rubioides, Crowea Saligna, Hibbertias and many other plants.[39]

However by 1882 these wildflower attractions were disappearing. *An Illustrated Guide to Sydney* lamented that in the country along Botany Road beyond Waterloo,

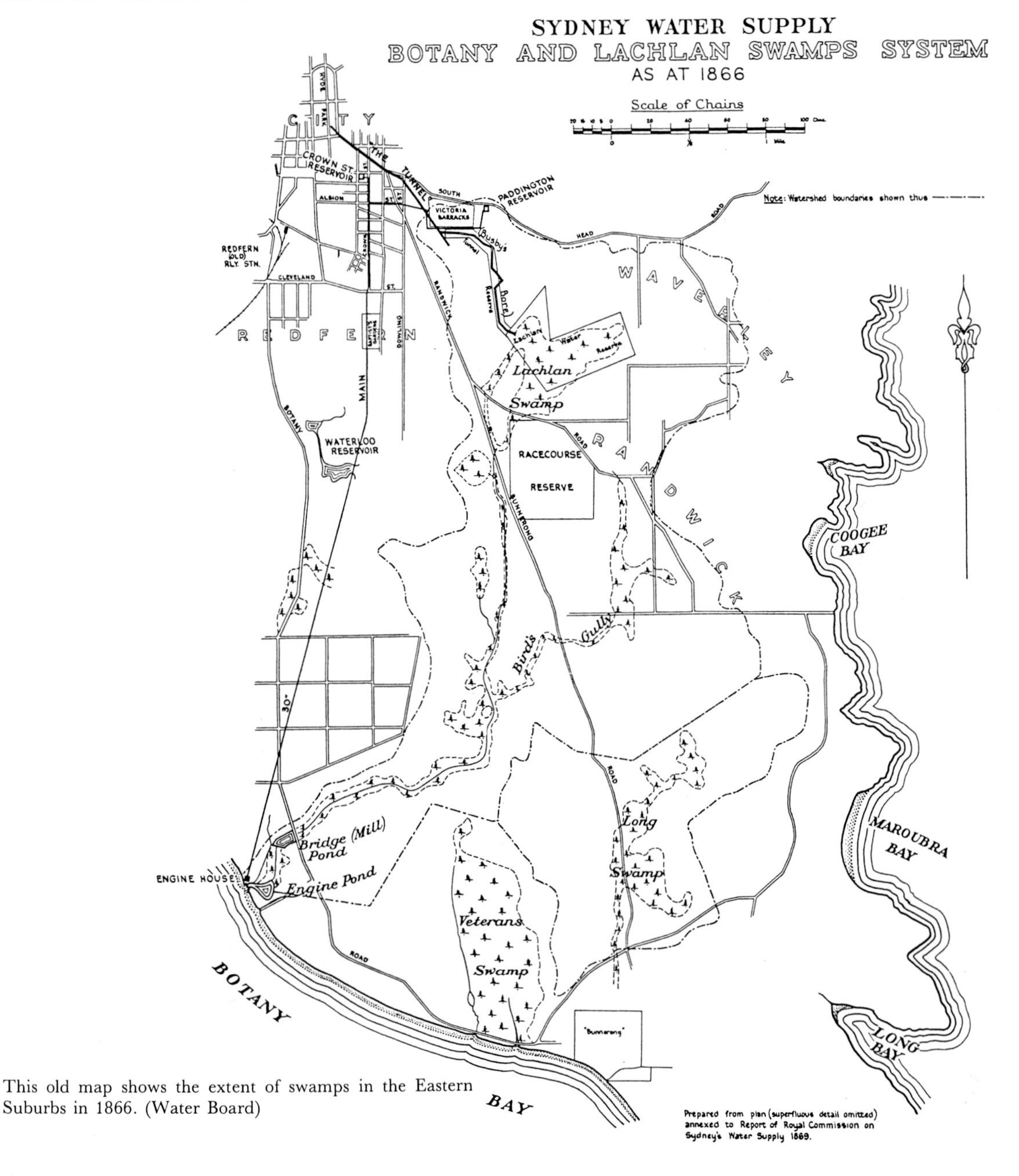

This old map shows the extent of swamps in the Eastern Suburbs in 1866. (Water Board)

Market gardens have usurped the place of the bright Epacris and the varied Boroniae. Vegetable life is but the counterpart of animal life — the uncultured must give way. There are, of course, a few specimens of wild flora, but in nothing like their old magnificence. Those who remember the road to Botany in years gone by are not surprised at the name given by the first discoverer [James Cook]... We know most of the wild flower regions of the colony, but none to compare in variety and richness with Botany, as it was.[40]

Extensive freshwater swamps were associated with the sand sheets either in the dune swales or along the major drainage lines. After his visit with Captain James Cook in 1770, Joseph Banks had indicated the suitability of Botany Bay for settlement, but Captain Arthur Phillip, on his arrival with the First Fleet in 1788, was not impressed, writing: 'the ground near it [Botany Bay], even in the higher parts, was in general damp and spongy'[1]. Similarly, after a nine-day survey in September 1789, Captain Watkin Tench wrote: 'We had passed through the country which the discoverers of Botany Bay extol as "some of the finest meadows in the world". These meadows, instead of grass, are covered with high coarse rushes, growing in a rotten spongy bog, into which we were plunged knee-deep at every step'[32].

The main wetlands were the Lachlan Swamps (now Centennial Park), the Botany Swamps (between Daceyville and Botany, in today's Lakes and Eastlakes Golf Courses), Long Swamp (now part of Heffron Park at Maroubra), and Veterans Swamp (at Banksmeadow). These wetlands were modified by water supply schemes or drained and filled for industrial or agricultural use. The Lachlan Swamps provided water for Sydney via Busby's Bore from 1827. Between 1860 and 1875 water storage capacity was increased by a series of dams and embankments on the Lachlan and Botany Swamps[41]. Remarkably little change has taken place in the configuration of these water bodies since that time. Unfortunately the permanent high water levels resulting from the dams have reduced the original wetland diversity, though they are still valuable natural areas. Before dam construction there would have been smaller, less permanent expanses of open water with patches of tall emergent sedges, fringed with zones of shorter sedges and occasional shrubs. *Eleocharis sphacelata* would have been a characteristic species, as it is today, and in the shallower water, *Baumea articulata, Baumea rubiginosa* and *Juncus* species would have been abundant. There is some *Typha* present now but this may be a response to today's higher nutrient levels. There does not appear to be any *Phragmites* now, though this species occurred around the estuary of

The Botany Swamps were dammed in the 1870s for Sydney's water supply but still provide an important habitat for waterplants and birds. Here is *Eleocharis sphacelata* at Eastlakes Golf Course. (1984)

Xanthorrhoea resinosa and a few shrubs cling to a sandhill in a corner of Eastlakes Golf Course. Except for a few tiny fragments like this, the colourful Eastern Suburbs Banksia Scrub that once stretched from here to the city has gone. (1984)

'Coogee Bay from Randwick Hill' in the 1870s showing heath and scrub on rocky outcrops in the foreground with woodland in the valley leading down to the Bay. (Government Printing Office Collection, State Library of New South Wales)

Xanthorrhoea still dominates the sand dunes near Maroubra Junction in the 1920s as a 'P' class tram passes the newly constructed houses. (Government Printing Office Collection, State Library of New South Wales)

the Cooks River and may have extended into the lower brackish swamps. Shrub species would once have been present, though only a few can be found today. The most conspicuous is the paperbark *Melaleuca quinquenervia,* which may occur as a small or large tree. There is a healthy patch of small trees in the Mill Stream section of the Botany Swamps, and a group of larger trees in the Lachlan Swamp at Centennial Park. Native Broom, *Viminaria juncea,* was common. 'In the Botany swamps the *Viminaria* frequently forms large shrubberies which are occasionally devastated by fire,' wrote Arthur Hamilton in 1918[42]. It may still be found occasionally there today. Other shrubs were *Callistemon citrinus* and *Callistemon linearis, Leptospermum juniperinum* and *Kunzea ambigua.*

In contrast to the sandy or peaty nature of the other swamp systems, Long Swamp near Malabar and Veterans Swamp at Banksmeadow appear to have been more fertile and were developed for market gardens in the nineteenth century; some of these market gardens still exist. The vegetation here was probably low forest of the paperbarks *Melaleuca ericifolia* and *Melaleuca linariifolia* and Swamp Mahogany, *Eucalyptus robusta,* with a grassy and herbaceous understorey.

The estuarine swamps of the Cooks River were obliterated by the upgrading of Kingsford Smith Airport in the 1940s, when the course of the river was redirected. Before this there were extensive mangroves, mainly Grey Mangrove, *Avicennia marina* with some clumps of *Aegiceras corniculatum,* along the Cooks River itself, and saltmarsh and dry salt plain, with Samphire, *Sarcocornia quinqueflora,* patches of *Juncus kraussii,* and the salt couches *Sporobolus virginicus* and *Zoysia pungens* on the margins of the marsh plain. Photos from 1919 indicate that the mats of Samphire were extensive. Estuarine swamps along Sheas Creek were destroyed when it was channelled in the 1890s (see South Sydney).

There were still extensive saltmarshes with *Sarcocornia quinqueflora* on the lower Cooks River in 1919. (Linnean Society of New South Wales)

Botany's low-lying barren sandy country was of limited value for agriculture, but with easy access to the city and port, was suitable for industry. The ready availability of water was important and the earliest industry, Simeon Lord's wool washery, was located on the Botany Swamps. Other industrial development took place late in the nineteenth century, using the plentiful supply of groundwater under the sand. The last remaining natural areas are now the Botany Swamp wetlands in the Eastlakes Golf Course, and tiny patches of Grass Trees, *Xanthorrhoea resinosa,* and heath plants on the golf course sand hills near Wentworth Avenue and at Banksmeadow Public School.

26 Randwick

Randwick occupies the eastern side of the Eastern Suburbs from Clovelly to La Perouse. It was first explored by James Cook during the stay at Botany Bay in 1770, when he 'made an excursion of 3 or 4 miles [about 5–6 km] into the Country or rather along the Sea Coast. We met with nothing remarkable, great part of the Country for some distance in land from the sea Coast is mostly a barren heath diversified with marshes and Morasses'[43]. Banks and Solander would have collected most of their specimens from the southern side of Botany Bay as most time was spent there, but at least one, *Bauera capitata*, is likely to have been collected at La Perouse as it has never been recorded naturally south of Botany Bay.

Cook's 'barren heath diversified with marshes and Morasses' is an appropriate description of Randwick's original vegetation for someone seeking agricultural land, but conveys little of the landscape's diversity and botanical richness.

Eastern Suburbs Banksia Scrub grew on the deep sand sheet but has now almost totally disappeared. The only place where its former diversity can be appreciated is at La Perouse near Jennifer Street, where there is a small sand dune with *Banksia aemula* and *Xanthorrhoea resinosa* scrub, associated with several small freshwater swamps. Well over 100 other species grow here. The area has been threatened by housing proposals, and will need protection against drainage changes to prevent deterioration in its floristic composition, particularly in the swamps.

A grove of Bangalays, *Eucalyptus botryoides*, and Angophoras, *Angophora costata*, on the foreshores of Yarra Bay in 1901. (Mitchell Library, State Library of New South Wales, with permission of Mrs Pillars, Randwick)

A century ago Centennial Park's 'wide expanse of hill and swamp' [46] had been laid bare, leaving scattered bushes and low-lying patches of freshwater swamp. (Mitchell Library, State Library of New South Wales)

Intensive grazing had eaten away the 'small scrub and various flowers' [39] characteristic of the sandsheet vegetation, leaving *Banksia serrata* bushes like these as sole survivors.

Elsewhere a few Banksias survive in the Bird Sanctuary enclosure at Centennial Park, and some naturally occurring plants of the rare *Allocasuarina diminuta* grow precariously in the median strip of Anzac Parade near Kingsford. The remaining 'marshes and Morasses' are the lakes of Centennial Park and that park's Lachlan Swamp Regeneration Area, a small remnant of the original swamp vegetation with Cutting Grass, *Gahnia sieberana,* and fern-banks beneath a dense forest of paperbark, *Melaleuca quinquenervia,* and evidently much as it was over 60 years ago when Arthur Hamilton wrote: 'These ferns [*Gleichenia dicarpa* and *Hypolepis muelleri*] may be observed engaged in such a competition in a peaty swamp in Centennial Park'[44].

Between South Head and Cape Banks the deep sand sheet that covers much of the Eastern Suburbs gives way to expose extensive sandstone rock

Xanthorrhoea resinosa and *Banksia aemula* are distinctive species of the very varied vegetation surviving on older sands at Jennifer Street, La Perouse. Well over 100 species can be found in this small Eastern Suburbs Banksia Scrub remnant. (1983)

Extensive areas of low heath on sandstone pavements still survive near Long Bay. (1978)

platforms and outcrops. Vegetation here was strongly influenced by exposure to ocean winds and salt spray, and by the degree of shelter and depths of soil available. On the rocky exposed headlands was low wind-pruned heathland with *Westringia fruticosa, Baeckea imbricata* and *Lomandra longifolia,* or taller heath and scrub with the predominant shrub species *Allocasuarina distyla, Banksia ericifolia, Leptospermum laevigatum* and *Melaleuca nodosa.* Smaller shrubs included *Woollsia pungens, Darwinia fascicularis, Epacris microphylla* and *Hakea dactyloides.* Remnants of this vegetation can still be seen at La Perouse and Long Bay.

Very little useful timber would have been obtained from the Eastern Suburbs, except possibly from the seaward facing valleys at Coogee, where Samuel Tree, a market gardener on the Waterloo Estate, recorded: 'Forty years ago (1820) I brought many a load of wood out of Coogee. . . what was called Coogee was a great gully where there was a great deal of timber, gum-trees, mahogany and other

A pleasant lagoon in the sandhills behind Bondi in the 1870s. Banksia scrub comes down the sandhills on the left to meet trees of *Melaleuca quinquenervia*, temporarily flooded. (Mitchell Library, State Library of New South Wales)

types'[45]. The 'gum-trees' may have been Smooth-barked Apple, *Angophora costata,* and the 'mahogany', Red Mahogany, *Eucalyptus resinifera,* or Bangalay, *Eucalyptus botryoides.* There is a popular belief that a sheltered sandstone gully with a number of large trees of Port Jackson Fig, *Ficus rubiginosa,* and a dense understorey of woody weeds in Glebe Gully, Randwick is a remnant patch of rainforest. Rather, it is probably a remnant of eucalypt forest; the few surviving native species are characteristic of moist sandstone gullies or creek lines, *Angophora costata, Ceratopetalum apetalum, Callicoma serratifolia, Breynia oblongifolia, Acmena smithii, Monotoca elliptica, Banksia integrifolia* and *Xylomelum pyriforme.*

Randwick developed as a salubrious suburban area in the late nineteenth century, when many grand houses were built. More intensive suburban subdivision followed in the early twentieth century, with development progressively taking place further from the city; Kensington, Kingsford, Maroubra, Matraville, Malabar and Chifley, over the next seventy years. Almost all this development replaced reasonably undisturbed native vegetation, and, together with recent developments at La Perouse and proposed developments at Long Bay, leaves only a tiny fraction of Randwick's original vegetation for future study and enjoyment.

27 Waverley

Waverley lies between the ocean coastline, from the Macquarie Lighthouse to Clovelly, and Old South Head Road, which generally follows the main highland divide between the ocean and the harbour. It is mainly a Hawkesbury Sandstone plateau cut in two by the low lying sand-filled valley between Bondi Beach and Rose Bay.

Behind the major beaches, particularly Bondi, were extensive deposits of unstable sand. Before the car parks and high rise there would have been low dunes with sand coloniser species, *Spinifex hirsutus* and *Festuca littoralis* nearest the ocean, and the typical coastal sand dune zonations through *Hibbertia scandens* and *Correa alba* to *Leptospermum laevigatum* and *Banksia integrifolia.* There is no record of any hind-dune rainforest associated with these beaches.

The sand between Bondi and Rose Bay appears to have been naturally unstable, and conspicuous to ships at sea. 'Went down to one of the lower coves and walked over to the Sand Hills which are given as a mark for a ship coming from the S. ward to know when they are near to Port Jackson, we found a good path over the neck of land not half an hours walk,' wrote Lieutenant Bradley in July 1788[9]. By the end of the nineteenth century the spread of housing along old South Head Road to Bondi and beyond was making drift-sand a problem. In 1900, referring to shifting sands in the vicinity of the tramway terminus at Bondi, J. H. Maiden, the director of the Botanic Gardens, reported:

> Here the sand filled up streets and obliterated fences, besides becoming a nuisance and an eyesore to the travelling public. Mr Cowdery levelled the sand and top-dressed it with a few inches of ashes from the tramway engines. The result should be witnessed by every person interested in this work of sand reclamation. A little couch grass was dibbled in here and there, and now we have a grassy lawn. Alongside, serving excellently for purposes of comparison, we have a neglected area, under the control of the municipality or park trust, as unsightly as the tramway portion is neat.[47]

The drift sand was finally stabilised by the cottage and residential flat development in the 1920s and '30s.

Extensive heath and scrub covered the sandstone plateaus. An idea of the variety of species present can be gained from Louisa Meredith's description of heath along Old South Head Road in 1844:

> The road, after descending the hill, turned to the left, through some sandy scrub, crowded with such exquisite flowers that to me it appeared one continued garden, and I walked for some distance, gathering handfuls of them — of the same plants that I had cherished in pots at home, or begged small sprays of in conservatories or greenhouses! I had whole boughs of the splendid metrosideros, a tall shrub, bearing flowers of the richest crimson, like a large bottle-brush [*Melaleuca hypericifolia*]; several varieties of the delicate epacris; different species of acacia, tea-tree and corraea, the brilliant 'Botany-Bay lily' [*Crinum pedunculatum*], and very many yet more lovely denizens of this interesting country, of which I know not even the name. One, most beautiful, was something like a small iris, of a pure ultra-marine blue, with smaller petals in the centre, most delicately pencilled; but ere I had gathered it five minutes, it had withered away, and I never could bring one home to make a drawing from.[21]

Undoubtedly Mrs Meredith was referring to a species of *Patersonia,* probably *Patersonia sericea.*

Bondi was predominantly a holiday destination during the nineteenth century though a considerable amount of suburban housing had been built along

Houses were begining to encroach on the extensive sand dunes behind Bondi Beach in 1919. The whole area was built on within 30 years. (Mitchell Library, State Library of New South Wales)

In 1909 a considerable amount of the heath and woodland vegetation of Vaucluse still remained. (Government Printing Office Collection, State Library of New South Wales)

Old South Head Road and adjacent areas by 1900, as a result of the efficient transport provided by the steam trams and later electric trams that dominated Eastern Suburbs transport until the 1960s. Suburban growth continued in the twentieth century with increasing higher population densities being achieved with multi-storey flats. Little consideration was given to retaining any original native vegetation and today it has virtually gone.

Blackbutts, *Eucalyptus pilularis*, along New South Head Road in the days when small farms alternated with grand mansions. (Mitchell Library, State Library of New South Wales)

28 Woollahra

Woollahra is the most northerly of the Eastern Suburbs municipalities and occupies the harbour foreshores from Rushcutters Bay to South Head, back as far as the main ridgeline along Old South Head Road. Except for the belt of low land extending from Rose Bay to the sand hills of Bondi, this is an area of sandstone slopes and gullies.

On the higher more exposed sites was heathland with a variety of shrubby species, like that described by Mrs Meredith along the Old South Head Road (see Waverley). This was particularly prominent at South Head and there are still significant remnants at The Gap.

On the more sheltered harbourside hillsides and gullies overlooking Port Jackson, taller woodland and forest would have been present, and may still be found at Nielsen Park, Vaucluse, now part of Sydney Harbour National Park, and in Cooper Park at Bondi, the head of the major gully draining to Double Bay. *Sydney Heads*, a painting by Eugene von Guerard showing the South Head peninsula in 1865 clearly shows the disposition of extensive heath and scrub on the exposed ridges and the taller woodland on the sheltered harbour side (see Figure 20). Mrs Meredith also describes this more wooded country:

> We drove back by a different road, nearer to the port, and less hilly, [presumably New South Head Road] but equally beautiful with that by which we came [Old South Head Road]. It led us through a moister-looking region, with more large trees, greener shrubs, and more luxuriant herbage, and commanding most lovely views, that appeared in succession like pictures seen through a natural framework of high white-stemmed gum-trees and tall acacias.[21]

Tree species here were Smooth-barked Apple, *Angophora costata,* Red Mahogany, *Eucalyptus resinifera,* Forest Red Gum, *Eucalyptus tereticornis,* Bangalay, *Eucalyptus botryoides,* and Scribbly Gum, *Eucalyptus haemastoma.*

The remaining bushland of the Eastern Suburbs has more than just scenic value. For example, a completely new she-oak species, *Allocasuarina portuensis*, was only recently discovered at Nielsen Park. Only ten plants of this species are known and particular care will need to be taken to ensure that it is not damaged inadvertently.

On the alluvial flats at the heads of bays such as Rushcutters Bay, Double Bay and Woolloomooloo Bay would have been forest with trees of Forest Red Gum, *Eucalyptus tereticornis,* Swamp Mahogany, *Eucalyptus robusta,* and Bangalay, *Eucalyptus botryoides,* with shrubs of *Kunzea ambigua, Leptospermum flavescens* and *Melaleuca ericifolia* and small 'rainforest-type' patches with Cabbage Palms, *Livistona australis,* used for constructing the first settlement's huts. 'I went down the harbour, with the master of the *Golden-Grove* victualler, to look for a cabbage tree as a covering for my hut,' recorded John White on the 21 July 1788[23]. This was probably the only vegetation approaching rainforest in the Eastern Suburbs.

At Rose Bay, on low sandy country, was a low swamp woodland of paperbarks, *Melaleuca quinquenervia,* that was evidently cleared in the late nineteenth century. Arthur Hamilton writes in 1919 that 'survivors of the extensive forest of these plants [of *Melaleuca quinquenervia*] originally in possession of the peaty flat stretching across to Rose Bay, are still preserved in paddocks and private gardens' [44]. Some trees still survive on the Royal Sydney Golf Club course. In the late nineteenth century a number of interesting plants were collected here including *Dodonaea falcata,* a species of the north-western slopes and Queensland with a couple of unusual occurrences near Sydney. At Rose Bay *Dodonaea* appears to have been an early casualty to development, the botanists Maiden and Betche writing in 1897: 'Now, we are sorry to say, *D. filifolia* [*falcata*] seems to be fast dying out in the Port Jackson district; hardly half a dozen plants could be found in 1896 in the same locality in which it abounded in 1883'[48]. Now extinct in the Eastern Suburbs, its only other Sydney occurrence is in the Castlereagh Woodlands near Penrith.

Blackbutts and Blue Gums grew large on shale-derived soils of the North Shore's ridge-tops. Many trees made a trip on transport like this to the nearest sawmill. (Hornsby Shire Library Local History Collection)

North of the Harbour

The municipalities and shires north of Sydney Harbour share the often spectacular and always interesting Hawkesbury Sandstone landscape. Timber-getting and farming cleared most of the shale-capped ridges, but large stretches of sandstone country were not disturbed by the nineteenth century agricultural expansion because of their poor sandy soils. Suburban development, limited until the opening of the Harbour Bridge in 1932, came relatively late. After World War II residential development escalated as a result of increasing car ownership. The car made lower density occupation possible, at the cost of a more widespread 'sprawl', accentuated by the increasing demand for suburbs set within scenic bushland. Today, mainly as a result of the steep topography, many bushland pockets and corridors of varying size and condition remain in these northern suburbs.

29 Baulkham Hills

In the early 1790s colonists began farming the alluvial and shale-derived soils along Toongabbie Creek, and north-east of Parramatta near the Burnside Homes. After his visit in 1792, David Burton, a gardener and surveyor acting as Superintendent of Convicts, reported of the latter area: 'for several miles to the northward and eastward of them, the ground is very excellent. It is a fine rich clammy light loam, from fifteen inches to two feet [40–60 cm] in depth'[49]. A government farm was established at Castle Hill in 1801, and by 1823 virtually all of the shale country had been allocated to colonists who were required, as a condition of their grant, to clear at least part of their land[50 51]. The effectiveness of clearing operations can be judged from John Lewin's painting (Figure 2).

This steep hilly Wianamatta Shale country that encloses an 'amphitheatre' cut out of the underlying Hawkesbury Sandstone by Darling Mills Creek and its tributaries, is occupied today by the suburbs of Baulkham Hills, Castle Hill and West Pennant Hills. Shale-covered ridges extend into this amphitheatre. On the higher rainfall eastern side, ridge-tops and upper hillslopes once grew Blue Gum High Forest of tall Blackbutt, *Eucalyptus pilularis*, and Sydney Blue Gum, *Eucalyptus saligna.* Most of this forest was logged, then gave way to farms and suburbs, but at West Pennant Hills in Cumberland State Forest

Blue Gum High Forest trees in the Hills district were still being cut at this sawmill at West Pennant Hills in the 1920s. (Courtesy of Nelson Blissett)

remnants can be seen. Here Blackbutts and Blue Gums grow with many High Forest understorey species — including the Forest Oak, *Allocasuarina torulosa,* the small tree *Glochidion ferdinandi,* shrubs *Clerodendrum tomentosum* and *Pittosporum revolutum,* and Sandpaper Fig, *Ficus coronata,* along the creeks. Additional specimen trees from other parts of the State have also been planted, so not all are locally representative[38 52].

West Baulkham Hills and the western parts of Castle Hill are drier and here the High Forest graded into Turpentine–Ironbark Forest. Turpentine, *Syncarpia glomulifera,* and Rough-barked Apple, *Angophora floribunda,* were common, with some Sydney Blue Gum and Blackbutt, in the gullies. Forest Red Gums, *Eucalyptus tereticornis,* grew particularly along the exposed steep western side of the Castle Hill ridge-line.

The valley of Darling Mills Creek, whose sandstone soils were considered too rough for farming, is able to support substantial trees because of enrichment by soil nutrients washed down over time from the surrounding shale ridge-tops. There are important natural areas here — Excelsior Reserve and Darling Mills State Forest Extension — where the influence of geology on vegetation can be seen. In Excelsior Reserve, as the track from Park Road down to Darling Mills Creek descends from the shale-capped ridge-top with the playing fields of the Ted Horwood Reserve, the vegetation changes from remnant Turpentine–Ironbark Forest — where Turpentine, Blackbutt, Forest Oak, Narrow-leaved Apple, *Angophora bakeri,* and White Stringybark, *Eucalyptus globoidea,* grow with clay-loving shrubs — to open-forest, typical of sandstone hillslopes. On the outcropping sandstone slopes, smooth-barked *Angophora costata,* Red Bloodwood, *Eucalyptus gummifera,* and Blackbutt are surrounded by a rich variety of sclerophyllous shurbs. At the bottom of the valley, Blackbutts and Blue Gums up to 25 m tall thrive in the moist nutrient-enriched soil. Also taking advantage of the better soil and moisture conditions along the creek are small trees common along creek lines in the sandstone country — *Elaeocarpus reticulatus, Callicoma serratifolia, Pittosporum undulatum, Lomatia myricoides, Tristaniopsis laurina* and a few Coachwoods, *Ceratopetalum apetalum.* The unfortunate effects of disturbance are also obvious here. Installation of the sewer main along the creek has allowed exotic weeds to gain a foothold, and in the nutrient-enriched creekside soil these will increase if left unchecked. Camphor Laurel, Privet, Crofton Weed, Blackberry, Fleabane, Cobblers Pegs and weedy grasses are among those present.

Baulkham Hills Shire includes more than just 'The Hills'. It extends from suburban Baulkham Hills, North Rocks and Carlingford, northwards fronting the Hawkesbury River from Cattai Creek to Wisemans Ferry. The interfaces between the shale of the Cumberland Plain, the sandstone of the Hornsby Plateau, and the floodplain of the Hawkesbury River provide a diversity of environments.

The north-eastern edge of the gently undulating country of the Cumberland Plain intrudes into Baulkham Hills Shire around Kellyville and along the Old Windsor Road. Here, on the deep clay soils were once woodlands with Grey Box, *Eucalyptus moluccana*, and Forest Red Gum, *Eucalyptus tereticornis*. Remnant trees and regrowth saplings can be seen in this area in narrow road verges and farm patches which have escaped clearing. There is a very important woodland area showing this type of vegetation at Longneck Lagoon, Pitt Town, in Hawkesbury Shire.

Between Castle Hill and the Hawkesbury River, the dissected sandstone of the Hornsby Plateau is the major landform. Most of the original Wianamatta Shale covering has gone, eroded in prehistoric times, leaving only narrow strips along the crests of the broader sandstone ridges. These shale-capped ridges would have supported Turpentine–Ironbark Forest, with some localised occurrences of Blue Gum High Forest trees on more favourable sites. Along Old Northern, Glenhaven, Annangrove, Kenthurst, Pitt Town and Porters Roads, the former shale forest has been cleared, leaving the sandstone vegetation on hillslopes and in the gullies. Forest on shale extending to Dural, Kenthurst, Annangrove and Glenorie has mostly been cleared for orchards and hobby farms

The O'Haras Creek catchment near Kenthurst contains a variety of sandstone plant communities, including uncommon ridge-top woodland and tall open-forest on the valley floor. (1990)

but a remnant in Ellerman Park near Round Corner, for example, gives some idea of how the long stretches of ridge-top forest appeared in earlier times.

Most of the Hawkesbury Sandstone country still retains its native vegetation. Woodland, with Scribbly Gums, *Eucalyptus haemastoma* and *racemosa*, Red and Yellow Bloodwoods, *Eucalyptus gummifera* and *Eucalyptus eximia*, Grey Gums, *Eucalyptus punctata,* Smooth-barked and Narrow-leaved Apples, *Angophora costata* and *Angophora bakeri,* covers the sandstone ridges and exposed hillsides as the Hornsby Plateau slopes gradually south-west to the Hawkesbury Valley. Sclerophyllous understorey shrubs include species of *Banksia, Grevillea, Hakea, Boronia, Leptospermum, Pultenaea* and their relatives. On broad plateaus shrublands with the spectacularly flowering Dwarf Apple, *Angophora hispida,* may be conspicuous.

On sheltered hillsides and in the sandstone gullies there is open-forest of Sydney Peppermint, *Eucalyptus piperita,* and *Angophora costata,* with Blackbutt and sometimes Sydney Blue Gum growing where soil fertility is better. At Kenthurst the O'Haras Creek valley provides an excellent example of the variety of plant communities which occupy the different topographic positions in the sandstone terrain, particularly as woodland remains on the undisturbed flat ridge-top land, one of few such areas to escape clearing. Yellow Bloodwoods and Narrow-leaved Apples grow on the steep dry upper sides of the valley here, with Grey Gums, Sydney Peppermints and Red Bloodwoods. Magnificent Sydney Blue Gums reach upwards more than 30 m from the fertile alluvial soil of the valley floor. These are regenerating well after logging 40 years ago. Blackbutts and Turpentines grow here too, along with many moisture-loving small trees, shrubs and climbers — *Acmena smithii, Ceratopetalum apetalum, Backhousia myrtifolia, Rapanea variabilis* and *Acacia implexa* to name but a few. Patches of reedy waterplants, *Phragmites australis, Typha orientalis* and *Eleocharis sphacelata,* grow along O'Haras Creek with occasional Swamp Mahogany trees, *Eucalyptus robusta,* and paperbarks, *Melaleuca linariifolia.* The fertile soil also provides favourable conditions for weeds that have gained a foothold in the disturbances caused by grazing and logging. The valley provides a spectacular contrast to the surrounding dry ridge-tops, though these of course, have their own interesting groups of species, including Woody Pear trees, *Xylomelum pyriforme,* and occasionally the rare Scaly-bark, *Eucalyptus squamosa* — elsewhere the continuing victims of Sydney's insatiable suburban sprawl.

'Without so much soil on the sandstone as would grow a cabbage,' recorded the *Australian* newspaper in 1827 of the Maroota district[53] further north. The former Maroota State Forest west of Old Northern Road encompasses steep, relatively untouched sandstone country. Although some sandstone ridges support trees indicative of clay influence — White Stringybark, Turpentine and Grey Gum — there are extensive sandstone rock platforms with shallow soil pockets. These provide habitat for the rare shrub species, *Kunzea rupestris,* though here the rapacious activities of the bush-rock industry have left their marks.

Better soil is encountered where the Maroota Ridge road reaches an outlier of Wianamatta Shale. Here the climate is drier, and instead of Turpentine-Ironbark Forest, the shale outlier supports woodlands of Forest Red Gum, *Eucalyptus tereticornis,* Ironbarks, *Eucalyptus crebra* and other species, like those on the Cumberland Plain to the south-west.

At the northern end of the shire the Hornsby Plateau rises to over 200 m, and the Hawkesbury River is confined within a deep valley. Dramatic cliffs intersected by small side-creek valleys are typical. At the foot of the cliffs, fertile alluvial flats flank the river and its tributary creeks, Cattai, Little Cattai, O'Haras, Blue Gum and Kellys Arm. Very little remains of their original floodplain levee-bank forest of *Eucalyptus tereticornis, Eucalyptus amplifolia,* Cabbage Gum, and rough-barked *Angophora floribunda*[54], cleared for orchards and farms in the early nineteenth century. Freshwater swamp vegetation occupied the backswamps and low-lying floodplain ground, trapped behind levees that formed as sea levels have risen during the last 20,000 years. These swamps have particular geomorphic and historical interest. Unfortunately most have been drained — those few that remain provide a valuable habitat for plants and waterbirds. Ducks, swamp hens and herons foraging amongst metre-high clumps of *Eleocharis sphacelata* growing in up to a metre of water, populate a very handsome swamp 2 km north of Lower Portland. Scrambling plants — *Persicaria praetermissa* and *Persicaria subsessile, Alternanthera denticulata* and the rush *Juncus usitatus* — consolidate the swamp edges while a scattering of white-flowering paperbarks, *Melaleuca linariifolia,* and a few Swamp Mahogany trees, *Eucalyptus robusta,* remain along the edges. Similar plant communities grow in Wheeney Lagoon, Broadwater Swamp and in smaller swamps extending several kilometres up slow-flowing Little Cattai Creek. As some of the best natural swamps still remaining along the Hawkesbury, these should be valued and managed carefully for their vegetation and wildlife values.

On very sheltered alluvial soils along the Hawkesbury, particularly in the narrow gullies of side creeks, was mesic forest with a rainforest-type understorey. A few small pockets persist beside the cliffs today. These plants provided a rich variety of resources for Aborigines who once lived here: fibrous inner bark of *Hibiscus heterophyllus* for string and fishing lines; succulent fruits of *Acmena smithii, Cissus hypoglauca, Eupomatia laurina, Ficus coronata, Ficus rubiginosa* and *Rubus hillii;* edible roots and tubers of *Cissus hypoglauca, Eustrephus latifolius* and *Geitonoplesium cymosum* vines and the herb *Geranium homeanum*; and wood of *Backhousia myrtifolia* for boomerangs, sanded smooth with leaves of the Sandpaper Fig, *Ficus coronata.* Together with the edible roots and tubers of nearby freshwater swamp plants, *Eleocharis sphacelata, Phragmites australis, Triglochin procera* and *Typha orientalis,* the floodplain provided plenty of food for Aborigines living near the river. Floodplain swamps and forest are a scarce and irreplaceable part of our heritage, both Aboriginal and European, and merit the highest protection.

30 Hornsby

Hornsby Shire occupies a large wedge-shaped section of the Hornsby Plateau's Hawkesbury Sandstone, between the Hawkesbury River, Cowan Creek and Old Northern Road. Its early European occupation was on two fronts. The Hawkesbury River was explored by Governor Phillip in 1788, and for the next half century limited farming and timber-cutting took place, though settlement here was constrained by the steep sandstone hillsides.

In the south, settlement of the fertile Wianamatta Shale country at Carlingford and Beecroft began soon afterwards. Tall Blue Gum High Forest extended along the high ridge now followed by the railway line from Epping to Hornsby, and across to West Pennant Hills. 'Very bad travelling, though in forest land,' reported the botanist George Caley in 1805 as he crossed through Beecroft towards the head of the Lane Cove River:

> The grass was very high, accompanied with a deal of thorn [*Bursaria spinosa*] and a species of *Platylobium.* The ground hilly, the gullies emptying their water to our right, probably into Lane Cove. The timber chiefly Blue Gum and She Oak. In some places small stones appeared, but taking the land in general it is of good quality, and more fit to be converted into arable than pasture ground.[55]

The Sydney Blue Gum, *Eucalyptus saligna,* Blackbutt, *Eucalyptus pilularis* and Grey Ironbark, *Eucalyptus paniculata,* of the Blue Gum High Forest probably grew to 30–40 m, and provided valuable timber. At Pennant Hills, for instance, the trees were described as: 'in general of an uncommonly large size, perhaps more so than in any other part of Cumberland, and therefore very advantageously situated so near a rapidly increasing town'[56].

Within the forest grew smaller trees of *Angophora floribunda, Acacia implexa, Pittosporum undulatum* and the Forest She-oak, *Allocasuarina torulosa,* noted by Caley. Shrubs of *Bursaria spinosa, Clerodendrum tomentosum, Pittosporum revolutum, Leucopogon juniperinus* and *Platylobium formosum,* vines including *Glycine, Clematis, Eustrephus latifolius* and *Tylophora barbata,* ground cover plants, *Geranium* and *Dichondra repens* as well as ground orchids and grasses, were all present.

The big trees were cut and the land cleared for farms and orchards; suburbs followed the development of the northern railway and today, no sizeable example of the impressive High Forest is left in Hornsby. The National Trust's Ludovic Blackwood Sanctuary at Beecroft preserves about two hectares, while Pennant Hills Park contains a small remnant (near George Street) where the transition from shale to sandstone vegetation can be seen. Individual large Blackbutts and Blue Gums can be seen along roadsides, and in small suburban parks such as Wollundry and Red Hill Parks in Pennant Hills, while in Kenley Park, Normanhurst, and The Village Green, Beecroft, large Grey Ironbarks also remain. In these small suburban parks the former understories of native shrubs and small trees have been replaced with mown grass.

On the drier Wianamatta Shale ridges further west, and where the shale is shallower as it gives way to the underlying sandstone, Turpentine–Ironbark Forest replaced the Blue Gum High Forest. This included trees of Turpentine, *Syncarpia glomulifera,* Grey Gum, *Eucalyptus punctata,* Grey Ironbark *Eucalyptus paniculata,* White Mahogany *Eucalyptus acmenoides,* Red Mahogany, *Eucalyptus resinifera,* White Stringybark, *Eucalyptus globoidea,* and rough-barked *Angophora floribunda.* There would have been such vegetation on the ridges from Dural to Arcadia and Galston, and along Old Northern Road to Glenorie, with outliers beyond. A sample can still be seen in Arcadia (in the north-eastern corner of

Sydney Blue Gums, *Eucalyptus saligna*, fall to the 'forest devil' in the Blue Gum High Forest on the shale soils at Epping. (Hornsby Shire Library Local History Collection)

Where creeks cut down into the sandstone, trees were smaller and less valuable for timber-getters, and bushland survived longer. At the turn of the century, Blackbutts still grew along this creek at Ray Road, Epping, and their saplings had established along the roadside. (Hornsby Shire Library Local History Collection)

Fagan Park). Individual *Syncarpia* and *Angophora floribunda* trees, indicative of shallower shale soil near the sandstone boundary, often persist along suburban roadsides.

Between the shale country and the Hawkesbury River the vast stretches of the Hornsby Plateau's sandstone country attracted little settlement until comparatively recently. The shallow soils are poor and the eucalypt woodland on ridge-tops and exposed slopes is typically dominated by twisted, characteristically marked Scribbly Gums, *Eucalyptus haemastoma* and *Eucalyptus racemosa,* and rough-barked Red Bloodwoods, *Eucalyptus gummifera,* with Old Man Banksias, *Banksia serrata*, and the shrubby papery-barked tea-tree, *Leptospermum attenuatum.* Other tree species include Narrow-leaved Stringybarks, *Eucalyptus sparsifolia,* Grey Gums, *Eucalyptus punctata,* and Yellow Bloodwoods, *Eucalyptus eximia.* Where soil is shallow or poorly drained, heaths or sedge swamps replace the woodland. Dense thickets of *Banksia ericifolia, Hakea teretifolia, Angophora hispida* or *Allocasuarina distyla* are common in the drier heaths. 'In many places the ground was bare of trees, but was thickly covered with *Metrosideros hispida* [*Angophora hispida*], whose stubbornness wore our clothes very much,' reported Caley in 1805 of such scrub on ridges in the upper Lane Cove valley[55]. Sheltered slopes and gullies have taller open-forest, with Sydney Peppermint, *Eucalyptus piperita*, and smooth-barked *Angophora costata* as the dominant trees, and the smaller *Allocasuarina littoralis, Elaeocarpus reticulatus* and *Ceratopetalum gummiferum.* Rocky creeks are lined with small trees — Water Gum, *Tristaniopsis laurina,* and Black Wattle, *Callicoma serratifolia,* while Coachwood, *Ceratopetalum apetalum,* grows along creeks with sufficient moist fertile soil. The dominant or most abundant species are only a small fraction of the total number present. In the Elouera section of Berowra Valley Bushland Park, for example, at least 420 native species have been identified. Some parks may feature particular species or unusually dense stands, such as the *Allocasuarina* in Florence Cotton Park at Hornsby.

Sandstone landscapes are conserved in Ku-ring-gai Chase and Marramarra National Parks, Muogamarra Nature Reserve, Pennant Hills Park and the Berowra Valley Bushland Park (which includes the former Elouera Bushland Natural Park)[57–61]. Plant communities on steep hillsides and gullies are particularly well represented in the local suburban reserves, but virtually no shale outcrops (apart from small shale lenses within the sandstone) are included, and plant communities near

The National Trust's Blackwood Reserve at Beecroft contains an important remnant of Blue Gum High Forest. (J. Plaza, RBG, 1990)

A stand of Turpentine–Ironbark Forest survives at Fagan Park, Arcadia. (1975)

Most of the bushland remaining in Hornsby today is on sandstone hillslopes, where open-forest is the characteristic type of vegetation. Here *Angophora costata* grows above a diverse array of shrubs in Pennant Hills Park. (J. Plaza, RBG, 1990)

shale/sandstone boundaries are poorly conserved. An example of these is in Thornleigh Park (near Handley and Dawson Avenues) at Thornleigh. Sandstone ridge-tops close to shale sites may contain a thin, often iron-rich remnant layer, from the base of the shale sequences. Those sites often have a characteristic flora including species which have become rare or vulnerable, because most similar flat ridge-tops have been cleared. Tunks Ridge in Berowra Valley Bushland Park is one of the few areas like this remaining undeveloped; the rare Heart-leaved Stringybark, *Eucalyptus camfieldii,* occurs here, along with several other uncommon species: Scaly-bark, *Eucalyptus squamosa,* Whip-stick Ash, *Eucalyptus multicaulis,* and *Darwinia biflora.* Pennant Hills Park included some areas of similar topography known to contain *Eucalyptus squamosa* and *Darwinia biflora*, but most of these have been cleared for sports fields.

Soils formed on the underlying Narrabeen Group shales and sandstones are more fertile than those of the Hawkesbury Sandstone. Where the Narrabeen Group outcrops along the banks of the Hawkesbury River and the lower reaches of Cowan and Berowra Creeks, open-forest with a different mixture of tree species and grassier understorey can be seen on the lower hillslopes. *Angophora floribunda, Allocasuarina torulosa* and *Eucalyptus punctata* are common. *Eucalyptus paniculata, Eucalyptus botryoides* and *Eucalyptus umbra* also occur. The soft felty leaves of *Astrotricha floccosa* may be conspicuous in the open shrub layer, particularly after a bushfire. Fringing mangroves and associated estuarine wetland communities are extensive in the lower reaches of Marramarra Creek.

The Hornsby diatreme, the largest in the Sydney metropolitan area, is an outcrop of volcanic breccia that forms the floor of Old Man Valley, west of Hornsby. The fertile soil previously supported a distinctive tall *Eucalyptus saligna* forest, but logging, farming and quarrying have removed most. A stand of *Eucalyptus saligna* can be seen on Old Mans Creek near the end of Rosemead Road; only part is protected in the adjacent Berowra Valley Bushland Park.

Much of Hornsby's rugged sandstone terrain remained undisturbed until after World War II, when the increasing availability of the car and improved building technology made steeper, more remote sites available for housing. As a result, bushland on ridge-tops and upper slopes has been almost totally destroyed, the bush remaining only where it is virtually impossible to build, and along steep gullies which have become drainage lines. Virtually every catchment system includes some suburban development, stormwater run-off from which contains silt and nutrients. These promote weed invasion of the sandstone gullies, and the effects are evident — in the older catchments such as Devlins Creek at Epping there is considerable invasion of Privet, *Ligustrum,* and many other weeds. In newer areas such as Mt Colah and Berowra, such invasion is beginning, and the consequences appear inevitable. Bushland is a significant component of Hornsby's environment, and the council has a particular responsibility to ensure its bushland management programme puts sound principles into practice.

31 Hunters Hill

Looking out over the Harbour from Kellys Bush on the Hunters Hill peninsula, it is easy for us to imagine Captain John Hunter's boat as he first explored the Parramatta River in 1788, passing the rugged slopes covered with shrubs and twisted *Angophora* trees, and startling Aborigines as they fished near rocky headlands and collected mussels among the mangroves. 'The Natives all fled in their Canoes as far & as fast as they could,' reported Lieutenant Bradley[9].

Road construction in the 1930s still involved a lot of manpower, and the sandstone country of the North Shore provided considerable obstacles. The development of diesel-engined earthmoving equipment after World War II led to rapid development of much of the more rugged parts. (Government Printing Office Collection, State Library of New South Wales)

The rocky peninsula of Hunters Hill has been a desirable residential address for well over a century. The sandstone which provides so much of its visual appeal made it useless for agriculture and, rather than wholesale clearing, the bushland suffered piecemeal destruction as houses proliferated on choice sites. Today, although the peninsula has many trees resulting from street and garden retention and planting, only a few small patches of natural bushland remain.

Kellys Bush is the largest bushland reserve on the peninsula, and provides access to *Kunzea ambigua* scrub, Sydney Peppermint, *Eucalyptus piperita,* and Red Bloodwood, *Eucalyptus gummifera*, open-forest characteristic of sandstone slopes, and dense foreshore thickets of *Ficus rubiginosa,* Port Jackson Fig, *Banksia integrifolia,* Coast Banksia, *Elaeocarpus reticulatus,* Blueberry Ash, *Acacia longifolia,* Sydney Golden Wattle, and *Pittosporum undulatum.* Many trees were cut to fuel Kelly's smelter, established in 1892. Local residents and concerned trade unionists were instrumental in saving this bush, when in 1971

Boronia Park in Hunters Hill contains a diversity of habitats; open-forest on sandstone slopes and tiny remnants of Turpentine–Ironbark Forest on shale survive above this pool on the edge of the estuarine fringe of mangroves along the Lane Cove River. (J. Plaza, RBG, 1990)

the first 'Green Bans' were imposed, preventing destruction of this waterfront bushland for housing, and earning Kellys Bush a significant place in conservation history[62].

Open-forest of sandstone slopes can be found in most of Hunters Hill's bushland reserves, though their small size means many parks have been subject to weed invasion, as at Gladesville Reserve, Betts Park and the Ferdinand Street Reserve, for example[63]. Perhaps Boronia Park is the best place to gain an idea of Hunters Hill's pre-European landscape, as it has the largest intact stretch of bushland. Its sandstone slopes carry characteristic open-forest of *Eucalyptus piperita, Eucalyptus gummifera* and *Angophora costata,* with a great variety of understorey shrubs and small trees. Towards the top of the slope is an area of *Kunzea ambigua* shrubland, while Blackbutt, *Eucalyptus pilularis,* grows on the more sheltered aspects, and a small creek tumbles down a steep waterfall to form a pool beside Grey

Mangroves along the river's edge. One can imagine this providing a suitable stopping-place for Aborigines collecting shellfish, and nearby middens have confirmed their occupation along the river here.

Extensive stands of the Grey Mangrove, *Avicennia marina,* occur along Buffalo Creek, although there has been some recent landfill. Along the Lane Cove River the narrow band of fringing *Avicennia* widens beside the southern section of Boronia Park. On the Parramatta River, mangrove stands fringing Gladesville Reserve and Tarban Creek are smaller. Ferdinand Street Reserve on the northern side of the peninsula has a small area of saltmarsh, but more can be seen in Buffalo Creek Reserve; here the original saltmarsh area has been reduced by landfill, and recently mangroves have been invading the remaining saltmarsh area[69 74].

Turpentine-Ironbark Forest grew on shale in the vicinity of St Joseph's College, marking the edge of extensive forest that covered much of the Ryde district. Although some of its species can be found in the open-forest of protected sandstone gullies, the Turpentine-Ironbark Forest's only remnant in Hunters Hill today is a small stand of *Syncarpia glomulifera,* with some understorey shrubs, near the entrance to Boronia Park. This important remnant contributes to the diversity of environments in Boronia Park, and is deserving of rehabilitation and careful management.

Tall open-forest of Blackbutt, *Eucalyptus pilularis*, Grey Ironbark, *Eucalyptus paniculata,* and *Angophora costata* with a shrubby understorey in a drier part of Dalrymple Hay Nature Reserve at St Ives. (1974)

32 Ku-ring-gai

Sandstone hillsides sloping steeply down to Middle Harbour Creek were the Europeans' first views of Ku-ring-gai. On 16 April 1788 Governor Phillip, with a small exploratory party, camped at 'a steep valley where the flowing tide ceased and a freshwater stream commenced'. After 'a long and fatiguing march' from the coast, the party seems to have been less than impressed with the vegetation of the beautiful rocky hillslopes and sandy creek flats. 'Here, in the most desert, wild, and solitary seclusion that the imagination can form any idea of, we took up our abode for the night,' and 'washed our shirts and stockings,' wrote Surgeon-General John White[23]. Today the site, 'Bungaroo' in Davidson Park State Recreation Area near East Killara, still gives visitors an impression of remoteness and seclusion, its creekside shrubs — Grey Myrtle, *Backhousia myrtifolia;* Scrub Beefwood, *Stenocarpus salignus* (a relative of the firewheel tree); *Lomatia myricoides,* and a mint bush, *Prostanthera denticulata* — providing a welcome contrast to the Azaleas and Liquidambars of the surrounding suburban gardens.

From Middle Harbour Creek, Phillip's party struck westward, through land 'covered with an endless wood'; this was evidently the Blue Gum High Forest on the Wianamatta Shale of the main ridge, probably around Pymble. John White noted 'the land here. . . was better than the parts which we had already explored'. But being unable 'to penetrate through this immense forest', they returned. The differences between vegetation on the sandstone hillsides, 'rather high and rocky, and the soil arid, parched and inhospitable', and the 'immense forest' on the shale-covered ridgetops were evident.

Ku-ring-gai Municipality straddles the broad ridge followed by the Pacific Highway, between upper Middle Harbour and the upper Lane Cove River. From Roseville to Wahroonga the central spine of the ridge is covered with fertile clay soils developed from the Wianamatta Shale. Here, with the highest rainfall in Sydney (Pymble receives 1,444 mm per annum), were magnificent stands of the Blue

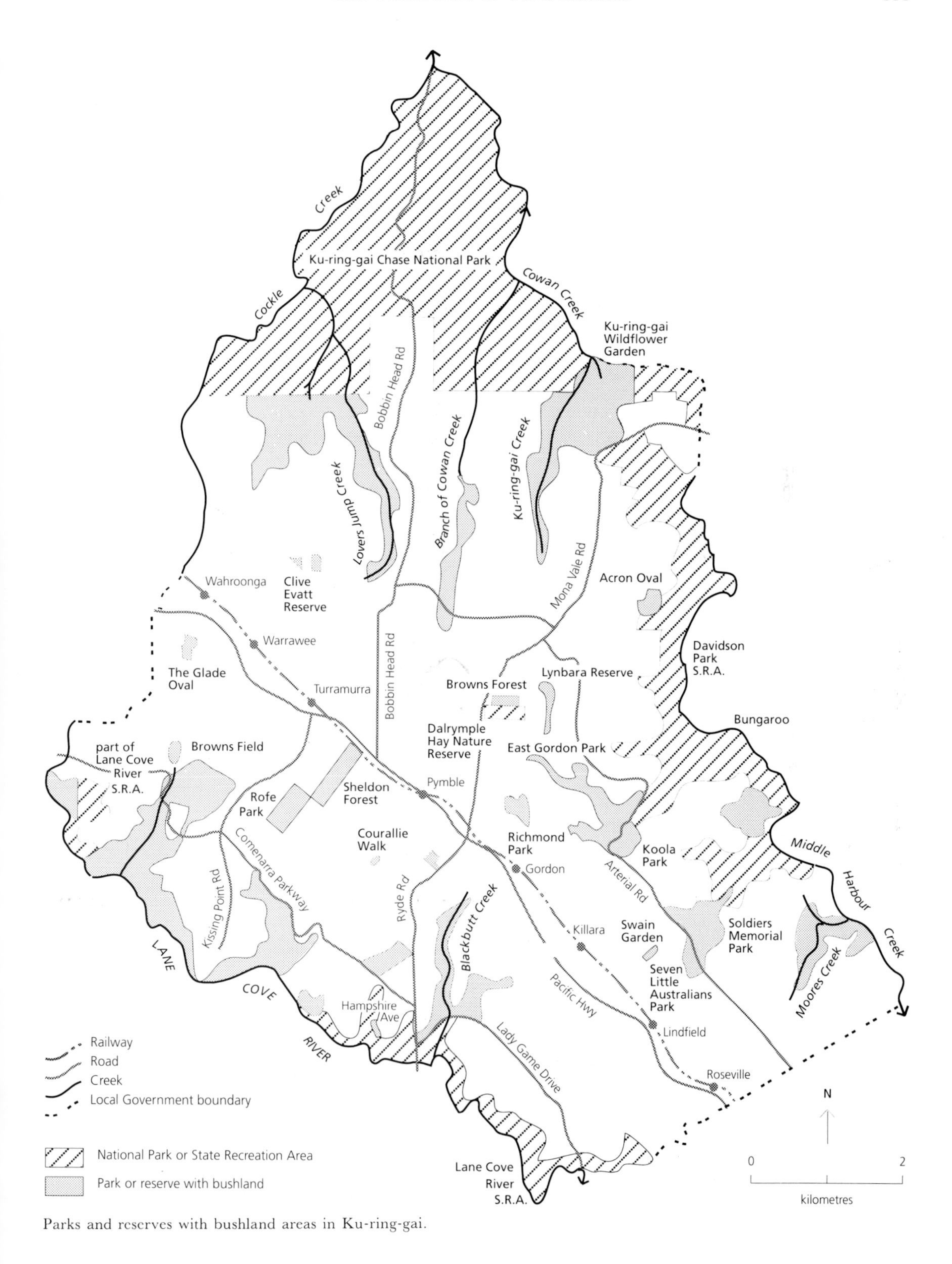

Parks and reserves with bushland areas in Ku-ring-gai.

Gum High Forest. This forest, John White's 'endless wood', extended in a band up to two kilometres wide along today's Pacific Highway from Roseville to Gordon, broadening north of Pymble, covering the ridge-line now followed by Mona Vale Road and the higher parts of St Ives and North Turramurra.

The tall straight trees of Sydney Blue Gum, *Eucalyptus saligna,* and Blackbutt, *Eucalyptus pilularis,* soon attracted timbergetters. In 1805 the botanist George Caley described the North Shore timber as being very suitable for building Sydney town[55]. One Wahroonga eucalypt was found to have a girth of 11 metres. Soon the forest resounded to the strains of bullock team and axe. Bullock drays and timber jinkers carted logs down Fiddens Wharf or Grosvenor Roads to the Lane Cove River, from where they were punted downstream to Sydney. River transport of timber continued into the 1870s. *Eucalyptus pilularis* was sought for general construction timber; *Eucalyptus saligna* for beams, floorboards, wheel rims, and ship-building; *Eucalyptus resinifera,* Red Mahogany, for furniture and ship framing; the natural preservative oils in Turpentine, *Syncarpia glomulifera,* made it ideal for wharves and construction work; Grey Ironbark, *Eucalyptus paniculata,* and Red Bloodwood, *Eucalyptus gummifera*, were used for fencing; and Forest Oak, *Allocasuarina torulosa,* was used in cabinet-making, and split into shingles. In 1813, for example, Governor Macquarie instructed the government sawpit on the Lane Cove River to supply, each week, 4,000 shingles or 600 palings per three-man gang and 450 feet (150 m) of sawn timber from each two man team[68]. After the tallest trees had been cut, the forests supplied Sydney residents with firewood for many years. Orchards were established on cleared forest land from as early as 1826.

Only tiny remnants of the ridge-top High Forest exist today. The largest, in the Dalrymple-Hay Nature Reserve and the adjoining Browns Forest on Mona Vale Road at St Ives, contains less than 16 ha of the forest that once covered about 40% of the municipality. Other remnants can be seen in the Clive Evatt Reserve (1.7 ha) in Wahroonga, and Sheldon Forest (7 ha) at Pymble[64]. On the eastern side of Fox Valley Road about 10 ha of forest owned by the Wahroonga Sanitorium contains Grey Ironbarks, amongst Red Mahoganies, Blackbutts and Blue Gums.

Trees of the broad ridge-top High Forest may also grow in sheltered valleys, where fertile soil has accumulated; for example, Blackbutt and Blue Gum trees grow where tributary creeks join the upper Lane Cove River. Turpentines and Blackbutts can be seen growing together in the headwaters of Cowan Creek, adjoining Bedford Road at North Turramurra; in Richmond Park, Gordon; on the steep gully slopes of High Ridge Creek, East Gordon; in Lynbara Reserve, St Ives; and in the lower part of Sheldon Forest. On sheltered lower slopes along the Lane Cove River, Turpentines and Blackbutts are joined by Sydney Peppermints and Red Mahoganies.

On the transition from shale to sandstone soils, the composition of the forest changed, with different blends of species depending on local conditions and vegetation history. Trees of *Syncarpia, Eucalyptus globoidea,* White Stringybark, and the smooth-barked *Angophora costata* grew on shallower shale soil closer to the sandstone boundary.

The poor sandy soils of the sandstone country provide different conditions for plant growth. On ridge-tops here was woodland with Scribbly Gums, *Eucalyptus haemastoma* and *Eucalyptus racemosa,* and Red Bloodwoods, *Eucalyptus gummifera,* with a diverse understorey of colourful sclerophyllous shrubs, including *Banksia serrata, Leptospermum attenuatum, Lambertia formosa,* species of *Grevillea, Boronia, Acacia* and many more. Where the soil was very shallow, woodland was replaced by heath and scrub up to 4 m tall, with dense thickets of tall shrubby *Banksia ericifolia, Allocasuarina distyla, Hakea teretifolia, Angophora hispida* and *Kunzea ambigua.* Where drainage was impeded, there was low sedgeland, with small shrubs such as *Kunzea capitata* and a variety of sedges. Pockets of soil on flat rocky outcrops supported dwarf shrubs including *Baeckea brevifolia, Kunzea capitata, Calytrix tetragona, Darwinia fascicularis,* and sedges. Dense scrub of *Banksia ericifolia* still occurs on ridge-tops behind Turramurra High School and near the end of Canoon Road, in the South Turramurra Bushland.[65] Except for other small areas along the eastern bank of the Lane Cove River, and in Davidson Park and Lane Cove River State Recreation Areas, most ridge-tops have been occupied by housing. Gone are thickets of *Boronia floribunda* 2 m tall that once hid smooth *Angophora* trunks west of Turramurra's Rofe Park — one example of the ridge-top bushland's many delights now lost to us.

Steeper sandstone slopes are more likely to have remained wooded. Here *Angophora costata* commonly grows with *Eucalyptus piperita,* Sydney Peppermint, and *Eucalyptus gummifera,* together with small trees of Christmas Bush, *Ceratopetalum gummiferum* and *Elaeocarpus reticulatus,* the Blueberry Ash. This community occurs quite extensively on hillsides along the Lane Cove River valley[66 67].

Small rocky creeks also survive though their vegetation has often been altered by nutrients and weed growth. When undisturbed they characteristically have small trees of *Acacia irrorata, Callicoma serratifolia, Lomatia myricoides, Backhousia myrtifolia,* or *Tristaniopsis laurina.* Coachwood, *Ceratopetalum apetalum,* grows along the upper Lane Cove River and along many creek lines, and can be seen in most small reserves that include the lower reaches of watercourses.

One particular valley in Turramurra, Browns Field, is of special interest as it includes a small diatreme with fertile soil derived from volcanic breccia. Though the forest has been partly cleared for an oval, a small remnant of rainforest vegetation with dense Lillypillies, *Acmena smithii,* and some Sandpaper Figs, *Ficus coronata,* Cabbage Palms, *Livistona australis,* Scentless Rosewoods, *Synoum glandulosum,* Native Laurels, *Cryptocarya glaucescens,* as well as shrubs and vines, survives. Bipinnate-leaved *Acacia schinoides* grows prolifically along the creek below. Trampling by users of the oval has damaged understorey plants and helped proliferation of weeds in the rainforest. As the only vegetation of its kind in Ku-ring-gai, this hectare deserves the utmost protection.

Ku-ring-gai has always been seen as a leafy municipality, and the Council has been one of the first to take its bush management responsibilities seriously. Reports on the condition of all bushland reserves have been carried out and professional bush management and regeneration staff are employed.

33 Lane Cove

In February 1788, after first encountering Lane Cove's Aborigines, Lieutenant William Bradley reported:

> We saw several Natives, some sitting around a fire, others were just landing with their Canoes, the moment they perceived us they ran off in great confusion & hurry. . . We found Mussels on the Fire, others in their Canoes & some dropt between both; their fright was so great that they went off without taking their fishing lines, spears or any thing with them.[9]

The wooded slopes of the lower Lane Cove River near Riverview were virtually untouched in 1904. (Government Printing Office Collection, State Library of New South Wales)

The use of the shores and hillsides was soon to change. The river flats where Aborigines gathered their seafood, the moist gullies where the native blackberries *Rubus parvifolius* and *Rubus hillii* grew, and the fertile forested ridge line, soon attracted the European settlers. The tall eucalypts were soon put to the axe. 'I went this day to visit the Government Saw Pits at Lane Cove,' reported Governor Macquarie in 1810, 'the timber there is getting scarce. The saw pits must soon be moved to another place where timber is more abundant. The Stringy Bark and Blue Gum Trees are the best and fittest for Buildings & Floorings'[68]. They grew in the Blue Gum High Forest on the shale soils capping the higher parts of the municipality, and in moist gullies. The 'Blue Gum Trees' were *Eucalyptus saligna* and the 'Stringy Barks' probably included *Eucalyptus pilularis,* the Blackbutt, *Eucalyptus resinifera,* Red Mahogany and *Syncarpia glomulifera,* the Turpentine.

Originally the name Lane Cove denoted a large part of the north shore accessible from the river, and much of the timber passing through Governor Macquarie's 'Saw Pits at Lane Cove' would have come from Ku-ring-Gai and Willoughby. By 1895, when Lane Cove's present boundaries were delineated, its Blue Gum High Forest had long since gone; with the best trees cut for timber, the remaining forest was gradually cleared for farming.

Hawkesbury Sandstone slopes and gullies lead down to the Lane Cove River. The river provided access for the European settlers, but the steep rocky slopes and ridges proved uninviting for agricultural pursuits. The higher sandstone ridges, today's suburbs of Greenwich, Northwood, Longueville, Riverview, Linley Point and Lane Cove West, were covered in heaths and woodland of Scribbly Gum, *Eucalyptus haemastoma,* and Red Bloodwood, *Eucalyptus gummifera,* and a wealth of colourful understorey shrubs, including species of *Banksia, Grevillea, Hakea, Acacia, Leptospermum, Boronia* and members of the pea family. On the slopes was open-forest with the smooth-barked *Angophora costata* and fibrous-barked Sydney Peppermint, *Eucalyptus piperita.* These sites with their river vistas became sought-after home sites in the mid-twentieth century.

Today it is in the creek lines and on surrounding slopes, too steep for housing, that most bushland remains and along several creeks downwashed clay and deeper soils provide sheltered, nutrient-enriched conditions for taller open-forest with some of the same tree species that attracted timbergetters to the Blue Gum High Forest of the heights.

In 1788 the hillslopes along the Lane Cove River alternated with small estuarine swamps formed at the mouths of the side creeks on deposits of silt. Saltmarsh with the low succulent Samphire, *Sarcocornia quinqueflora,* and Salt Couch, *Sporobolus virginicus;* reed or rush swamps with *Phragmites australis* or *Juncus kraussii;* fringe forest of Swamp Oak, *Casuarina glauca,* and bands of paperbark, *Melaleuca linariifolia,* would have all been present. Old survey maps indicate that the now conspicuous Grey Mangrove tree, *Avicennia marina,* has increased in abundance since European settlement, particularly

Blackbutt stands still grew beside the Pacific Highway at St Leonards station in 1910. (Mitchell Library, State Library of New South Wales)

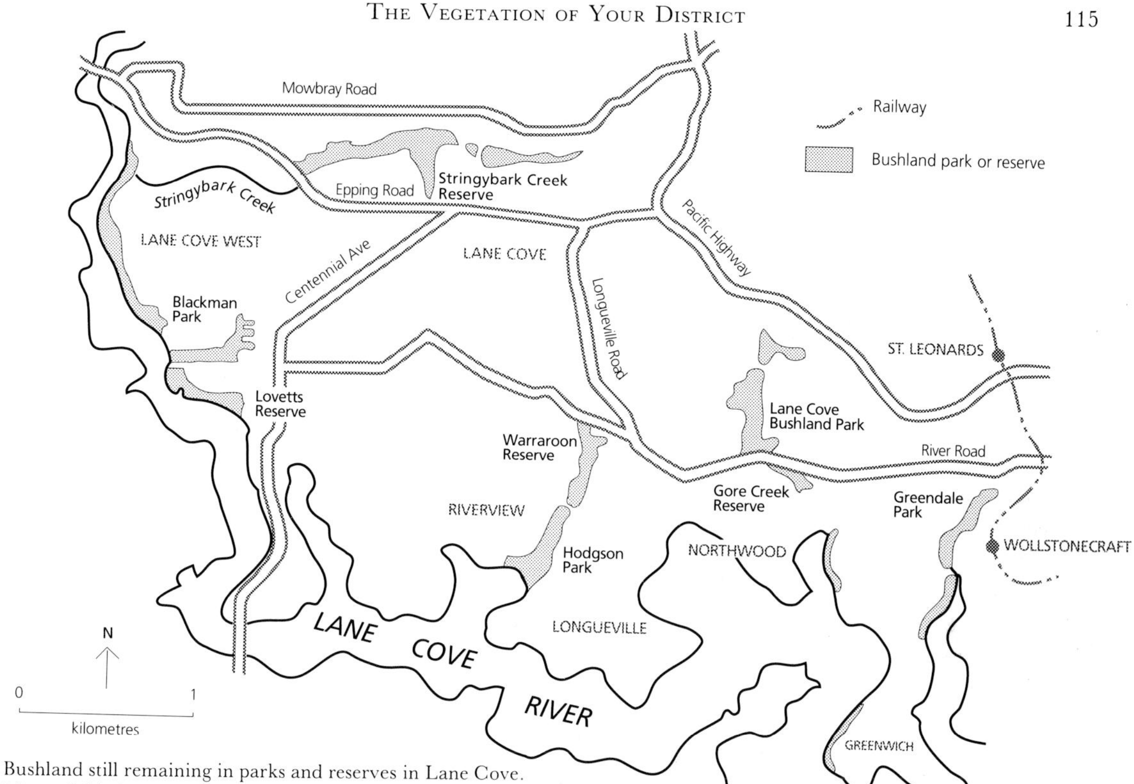

Bushland still remaining in parks and reserves in Lane Cove. Most of this is eucalypt woodland and open-forest on steep sandstone slopes, with trees of *Angophora costata* and *Eucalyptus piperita*. Mangroves, mainly *Avicennia marina*, fringe the Lane Cove River upstream from Figtree Bridge.

in the last 60 years, as housing and industrial development in the catchment has led to increasing sediment in the river[69 70]. However other wetland types have been depleted, by landfilling for industrial sites since the 1880s, and for ovals for recreation since the 1950s. Rarely is there a complete sequence of plant communities across the river flats today.

The Blue Gum High Forest has long gone, but remnants of Lane Cove's sandstone and wetland communities can still be seen. Stringybark Creek has the longest stretch of bushland remaining in the municipality. North of Epping Road, on the higher sandstone slopes, Sydney Peppermints, smooth barked *Angophora costata* and Stringybarks form open-forest. Closer to the creek there is a surprising amount of Black Wattle, *Callicoma serratifolia,* Blueberry Ash, *Elaeocarpus reticulatus*, large old Lillypillies, *Acmena smithii,* and some Coachwoods, *Ceratopetalum apetalum*, 10 m tall. The exotic privets *Ligustrum lucidum* and *Ligustrum sinense,* and Wandering Jew, *Tradescantia albiflora,* form some weedy patches, while further downstream the banks are heavily overgrown with Balloon Vine,

Grey Mangroves, *Avicennia marina*, grow vigorously and have colonised the increased sediment deposits that have built up in the Lane Cove River as a result of earthworks and clearing in the catchment. (1980)

Cardiospermum grandiflorum. Downstream from Epping Road the valley widens, and understorey shrubs become more dense and diverse. *Phebalium dentatum* in the *Boronia* family appears. Scribbly Gums and a few narrow-leaved *Angophora bakeri* trees at the top of the sandstone slopes are interspersed with small patches of *Kunzea amibigua* heath. Large figs, *Ficus rubiginosa,* hug vertical rock faces. Smooth-barked *Angophora* and the Sydney Peppermint, *Eucalyptus piperita* dominate the open-forest on the steep hillsides midslope, while the Blackbutt *Eucalyptus pilularis,* and the Turpentine, *Syncarpia glomulifera,* grow nearer the valley floor. The new link road, Sam Johnson Way, has removed much of the *Angophora bakeri,* which is now rare in Lane Cove. The steep road embankments have been planted with over 20 species of trees and shrubs native to the area, but the road cuts the natural corridor. Below the link road a large artificial freshwater lake and overgrown old ash dumps have replaced the complex of wetland communities which once occupied the lower Stringybark Creek valley floor.

Lane Cove Bushland Park, in the valley of Gore Creek, is another of the municipality's important reserves. The park includes the moist gully plants, *Trochocarpa laurina, Acrotriche divaricata, Morinda jasminoides* and the Filmy Fern *Hymenophyllum,* their growth aided by downwash of fertile shale soil from the ridge-line above at Gore Hill. An important remnant of the original vegetation, this forest was nearly cleared to enlarge the adjoining golf course. Timely action by local residents saved it.

In Warraroon Road Reserve, *Elaeocarpus reticulatus* and *Pittosporum undulatum* are abundant beneath large trees of *Angophora costata* along the upper reaches of the creek. The dense understorey includes *Bursaria spinosa,* usually a sign of soil fertility, and three species of *Cyathea,* while ground cover plants include *Dianella revoluta* and *Pteridium esculentum.* Saltmarsh and mangroves are still found lower down in Hodgson Park where the creek flows into Tambourine Bay. Gore Creek Reserve at Northwood is similar, though on a smaller scale, with *Angophora*-covered hillsides sloping down to small tidal flats with saltmarsh. Greendale Park at Wollstonecraft is larger; its lower reaches are quite swampy. Such areas were common where small creeks flowed into the harbour, but most have been filled to make sports fields (as at Burns Bay Oval).

Perhaps the best sequence of wetland communities remaining is found in Blackman Park, on a small riverside terrace along the Lane Cove River just south of Stringybark Creek. Tracks and boardwalks allow closer viewing. Along the river's edge are Grey Mangroves and Swamp Oaks. Landward of these, dense clumps of *Juncus* encircle a healthy little saltmarsh — with Samphire, *Sarcocornia quinqueflora,* the Salt Couch, *Zoysia macrantha,* and the herbs *Samolus repens* and *Suaeda australis.* Just below the edge of the drier sandstone hillside are paperbarks, *Melaleuca linariifolia,* Wild Spinach, *Tetragonia tetragonioides,* and the sedge *Baumea juncea.* The rocky slopes have typical sandstone open-forest, with Red Bloodwoods and the Red Mahoganies. The native ground-hugging conifer, *Podocarpus spinulosus,* and clumps of ferns following small drainage lines are of interest. Former wetland areas occupying the centre of Blackman Park — once rich sources of food for Aborigines — have been filled to make sports fields.

The bushland areas of Lane Cove contribute significantly to the quality of its suburban life, and Lane Cove Council was one of the first to recognise the need for serious management. Programs of bush regeneration and weeding, developed by the National Trust and now used by many Sydney councils, were adopted first here in the late 1970s.

34 Manly

Manly Municipality occupies the lower end of the Warringah Peninsula, between the ocean shoreline south of Queenscliff and Middle Harbour. The landscape is dominated by Hawkesbury Sandstone of moderate relief, overlain by sands along the beaches and silts associated with the coastal lagoons. The early development of Manly was as a tourist resort, with limited rural areas, and it was only after World War II that intensive residential development began. As a result Manly still has a considerable number of small remnants of bushland, as well as the most precious parts of Sydney Harbour National Park — North Head and Dobroyd Head. These two headlands, originally reserved for the military, contain the most extensive heath and scrub vegetation around Sydney Harbour, much of it in almost undisturbed condition. Dominated by shrubs of *Allocasuarina distyla* and *Banksia ericifolia,* this vegetation may be less than 0.5 or up to 3 m high, depending on the degree of exposure and its age since it was last burnt. Particular species are associated with localised conditions: for example, low, wind-pruned bushes of *Baeckea imbricata* and *Melaleuca armillaris* where there is direct exposure to salt spray; the fern *Gleichenia rupestris* in crevices along the cliff-lines; and the shrubs *Melaleuca hypericifolia* and *Hakea*

Manly Wharf in 1878 showing extensive areas of scrub and woodland on the sandstone peninsula leading to North Head. (Government Printing Office Collection, State Library of New South Wales)

Some very large remnant trees of *Melaleuca quinquenervia* in the 1890s, on the site of the present Manly Art Gallery. The size of the trees and their location is surprising as elsewhere around Sydney they were found in swamps as at Bondi and Botany. (Macleay Museum)

teretifolia where it is sheltered, but wetter underfoot. A few poorly drained creek lines have Button Grass sedge-swamp with *Gymnoschoenus sphaerocephalus* and the unmistakable *Banksia robur.* Heath on North Head, damaged by poorly-managed vehicular access and erosion, has been successfully rehabilitated.

Perched on the broad central summit of the sandstone headland of North Head is an interesting deposit of white leached sand with scrub of *Leptospermum laevigatum, Banksia aemula* and *Monotoca elliptica* that has been dissected by numerous roads. Parts of this have not been burned for more than 30 years, and changes are apparent in the vegetation as *Leptospermum* plants age, and die, to be replaced by *Pittosporum undulatum* and the exotic *Lantana camara.* There are also a number of dead and moribund plants of the rare *Eucalyptus camfieldii.* The role of periodic fire in maintaining this vegetation is evident. On sheltered sites away from the sea is woodland and open-forest with trees of Bangalay, *Eucalyptus botryoides,* Sydney Peppermint, *Eucalyptus piperita* and the smooth-barked *Angophora costata.* Similar vegetation was once common on the now residential areas of Seaforth, Balgowlah, Clontarf and Fairlight.

Sand dunes along the ocean beaches would have had typical bands of plants responding to salt spray and exposure; from the tough sand-binding *Spinifex hirsutus* on the fore-dunes, to shrubby Coastal Tea-tree, *Leptospermum laevigatum,* thickets on the crests, and patches of woodland or even rainforest scrub, on the sheltered lee sides. 'On the sheltered banks of Cabbage-tree Bay a limited "brush" association is established,' reports Arthur Hamilton in 1918. 'Two Native Grape vines, *Vitis* [now *Cissus*] *hypoglauca* F.v.M., and *V. Baudiniana* F.v.M., [now *Cissus antarctica*] are growing side by side — a frequent occurrence — their branches interlaced. *Pittosporum revolutum* Ait., occupies a more exposed position on the hillside'[42].

The dune vegetation disappeared during beach-side tourist development, to be replaced by the plantings, fashionable in the late nineteenth century, of elegant rows of Norfolk Island Pines, *Araucaria excelsa.* Sadly these too have now virtually disappeared, affected by detergents in wind-borne sea spray containing sewage effluent, that render the plant vulnerable to salt damage. Native vegetation still remains near Cabbage-tree Bay though the introduced *Erythrina* or Coral Tree predominates, planted before 1918 according to Hamilton: 'On the strand at the base of the hill a row of Coral Trees, *Erythrina indica* Lam., has been planted'.

A dairy on the low fertility sandstone soils at Manly in the 1920s. The photo shows the Sydney Peppermints, *Eucalyptus piperita*, stripped of their lower bark by cattle that have already denuded the ground of its shrubs and grasses.

Also gone is the natural vegetation around Manly Lagoon, which Governor Philip's exploratory party first met with in April 1788:

> The governor, anxious to acquire all the knowledge of the country in his power, forded the river in two places, and more than up to our waists in water, in hope of being able to avoid the thicket and swamps; but, notwithstanding all his perseverance, we were at length obliged to return and proceed along the sea-shore.[23]

The impenetrable 'thicket and swamp' would have included typical estuarine swamp forest with *Casuarina glauca* or, in less saline sites, *Eucalyptus robusta,* the Swamp Mahogany, and was similar to that still found further north around Narrabeen Lakes today.

35 Mosman

Mosman is a small municipality occupying Hawkesbury Sandstone ridges on the northern side of Port Jackson. The central spine, followed by Military, Spit and Bradleys Head Roads, is generally level and was cleared for houses in the nineteenth and early twentieth centuries. Along here, and on the more sandy soil extending down to the harbourside slopes was eucalypt woodland with trees of *Eucalyptus botryoides,* the Bangalay, Sydney Peppermint, *Eucalyptus piperita,* and the smooth-barked *Angophora costata.* Shallow sandy soils had woodland with Scribbly Gum, *Eucalyptus haemastoma;* Red Bloodwood, *Eucalyptus gummifera;* and *Banksia serrata.*

Mosman Bay in the 1870s when the slopes carried heathy eucalypt woodland. (Government Printing Office Collection, State Library of New South Wales)

In the late 1880s creeks flowing into Mosman Bay were unpolluted, and the bushland weed-free. Today such watercourses are choked with exotic weed species that have been encourged by the increasing levels of nutrients and silt. (Mitchell Library, State Library of New South Wales)

The name Bradley is stongly linked with Mosman's bushland. Bradleys Head, near Taronga Park Zoo, was named after William Bradley, First Lieutenant of First Fleet flagship, HMS *Sirius*. In detailed descriptions and individualistic watercolours[9], Bradley has left us in his journal a priceless record of Sydney Harbour's shores as they were in 1788. Now part of Sydney Harbour National Park, Bradleys Head retains woodland vegetation, similar to the 'Rocky Shore and thickwood on the Hills' described on Bradley's 1789 chart of nearby Mosman Bay. The wooded slopes of Mosman Bay, though still untouched nearly a century later, were to be lost as ferries and trams made Mosman's wooded shores a desirable residential address.

The Bradley Bushland Reserve (formerly part of Rawson Park) honours Elaine and Joan Bradley,

RED HONEYSUCKLE (*Banksia serrata*, LINN. F.), MOSMAN

Banksia serrata would have been common in woodland and heath on Mosman's sandstone ridge-tops.

Healthy open-forest with *Angophora costata* on the harbourside slopes of Bradleys Head probably looks much the same as when William Bradley visited here in 1788. Elsewhere on the headland eucalypts are suffering dieback in response to changed conditions. (J. Plaza, RBG, 1990)

local Mosman residents who set bushland management on a new course in the early 1970s by advocating the selective removal of exotic weed species and avoidance of unnecessary disturbance, rather than the bulldozing and broadscale spraying then currently practised[71 72]. Situated on Middle Head Road, the Reserve has an interesting remnant of the heath vegetation once common on exposed sandstone ridges around Sydney, but now largely replaced by housing. The characteristic large scrubby heath species *Banksia ericifolia, Kunzea ambigua* and *Allocasuarina distyla* are present, as well as the rare low-growing *Rulingia hermanniifolia,* seen only after a fire in a long unburnt area stimulated seed germination and growth.

Mosman is fortunate to have so much bushland, and particularly along its harbour foreshores. Such sites in other municipalities, particularly on the southern harbour foreshores and the lower Georges River, have been built out, but because they were required for military installations in the past, many of Mosman's bushland areas survived relatively unscathed to become part of Sydney Harbour National Park when no longer required for defence. On Middle Head, Chowder Head, Bradleys Head and in Ashton Park, winding bush paths lead past derelict military relics. Sydney Peppermint and *Angophora* canopies have regrown to obscure the sight-lines of the nineteenth century gunners. In many of these areas harbourside bushland is much as it was when the First Fleet sailed in, and a variety of *Acacia, Banksia, Leptospermum* and other shrub species can be seen. However, changes in drainage and fire frequency have allowed weed species such as Privet and Lantana to invade many of the wetter gullies, and there has been a considerable amount of recent localised deterioration in the eucalypt canopies, variously attributed to airborne pollution, sea spray, drainage changes, fire, lack of fire, and insect defoliation. Management will need to use a number of techniques involving fire, drainage control and bush weeding to maintain the diversity of native species in the long term.

36 North Sydney

North Sydney occupies the steep sandstone shores of the northern side of Sydney Harbour between Cremorne Point and Berrys Island, Wollstonecraft. Its early settlement was not rapid. In 1793 Captain John Hunter reported that 'the natives were employed in burning the grass on the north shore opposite to Sydney, in order to catch rats and other animals'[25].

The poor sandy soils discouraged farming. In 1806 James Milson, who farmed at today's Milsons Point, complained that his land appeared to be

Early artists used bushland at North Sydney as a foreground to pictures of Sydney Cove and generally stylised some of the most characteristic native species. *Xanthorrhoea* and *Casuarina* are clearly visible in this 1811 *View of the town of Sydney taken from Chiarabilly* by J.W. Lewin. (Mitchell Library, State Library of New South Wales)

nothing but rocks and stones, and there was not enough soil to grow anything[73]. The heights of North Sydney, however, appealed to artists, who climbed to favourite vantage points to paint panoramas of Sydney town across the water. In their foregrounds we can recognise many plants of the woodland, heath and open-forest that flourished in the sandy soils so disappointing to Milson — *Angophora, Allocasuarina, Xanthorrhoea.* As late as 1863 'from the water-front to Middle Harbour thick bush existed, with only one narrow track through, which is now Ben Boyd Road'[73]. The opening of the Harbour Bridge in 1932 and the high-rise growth of North Sydney since the 1960s have changed the landscape dramatically.

North Sydney's plant communities would have followed the vegetation patterns evident elsewhere on the North Shore — forest on the shale capping and woodland on the sandstone slopes. Back from the Harbour's shores, the Blue Gum High Forest covered the shale-topped ridge of Crows Nest and the higher parts of Wollstonecraft, Cammeray, North Sydney and Neutral Bay. Its tall Sydney Blue Gums, *Eucalyptus saligna,* and Blackbutts, *Eucalyptus pilularis,* 30 m or more in height, were cut for timber, and the land then farmed. On choice sites forest remnants provided the backdrop for some of the North Shore's earliest stately homes. No trace of this majestic forest remains.

Along the shoreline small tidal flats blocked the mouths of creeks in Balls Head Bay, Neutral Bay and Long Bay, but Gore Cove is probably the only place where these have not been destroyed by landfill. At Tunks and Primrose Parks, on Middle Harbour, landfill has drastically altered the original foreshores though the rocky sandstone slopes, more exposed than the south-facing slopes in the main harbour, still support eucalypt open-forest, albeit with considerable weed invasion.

Balls Head and Berry Island Reserves have also survived pressures for urban building and are more isolated from disturbance and weed sources than the Middle Harbour parks. On Balls Head, the vegetation has been altered in many places with plantings not native to the area, and with mown grass, but Berry Island allows us to see the open-forest that once covered North Sydney's sandstone foreshores. *Angophora costata* trees, with twisted pink trunks and branches, are abundant, together with Red Bloodwood, *Eucalyptus gummifera,* Sydney Peppermint, *Eucalyptus piperita,* Bangalay, *Eucalyptus botryoides,* Black She-oak, *Allocausarina littoralis,* and occasional Red Mahoganies, *Eucalyptus resinifera,* and

Berrys Island is much less disturbed than Balls Head, and provides an opportunity to retain harbourside vegetation in a natural state. There are no problems with stormwater intrusion, the only long-term management concern will be how to use fire to best advantage. (J. Plaza, RBG, 1990)

Grey Gums, *Eucalyptus punctata.* On the headland *Banksia integrifolia* is common, braving the salt spray, and large Port Jackson Figs, *Ficus rubiginosa,* cling to the lower sandstone slopes. *Glochidion ferdinandi,* Cheese Tree, and *Elaeocarpus reticulatus,* Blueberry Ash, are very common small trees in the open-forest, as is *Pittosporum undulatum.* Species characteristic of more fertile sheltered sites — *Notelaea longifolia, Dodonaea triquetra, Pittosporum revolutum, Polyscias sambucifolia, Clerodendrum tomentosum, Grevillea linearifolia* — are present, but so also are sclerophyllous shrubs, ground covers and grasses. The variety of plants is indicative of the varied habitats present in the irregular sandstone topography. Considering the proximity of the city centre, it is a delight to find these patches of native vegetation still remaining. Their survival requires constant vigilance and care. The death-knell of the native vegetation in the Cremorne Reserve on the eastern side of Shell Cove was finally sounded in the 1970s when unemployment relief teams, 'cleaning up', removed most vegetation, including natives, resulting in an overwhelming regrowth of weeds.

St Thomas's Church at North Sydney was built almost within a forest of tall Blackbutt, *Eucalyptus pilularis*, as this 1866 picture shows. (Mitchell Library, State Library of New South Wales)

Bernard Holterman's panorama of 1881 shows the woodlands and heath around the harbour giving way to housing. Balls Head, in spite of its imposing position, luckily missed out. Although the summit was cleared in the 1930s, and has been planted with many non-indigenous species, there are still patches of bush that should be allowed to regenerate. Unfortunately exotic species are being allowed to invade. (Mitchell Library, State Library of New South Wales)

This early view of 'Willoughby Falls' in Primrose Park, Cammeray, is now obscured by thick growth of Privet and Balloon Vine, *Cardiospermum grandiflorum*. (Mitchell Library, State Library of New South Wales)

37 Ryde

Ryde occupies the country between the Lane Cove and Parramatta Rivers as far north and west as Macquarie University, Eastwood and Meadowbank. The Hornsby Plateau here slopes to the south-west, and the rivers have cut through its Wianamatta Shale covering, to expose the Hawkesbury Sandstone in their valleys. Like other northern municipalities, the geology has the strongest influence on the vegetation, though climatic influences can also be inferred, particularly in the shale vegetation. Ryde's rainfall is intermediate between the higher rainfall of the North Shore ridge and the drier country west of Parramatta.

The vegetation of the shale country in areas of moderate rainfall (900–1,100 mm per annum) was mainly Turpentine-Ironbark Forest. In Ryde much of this forest was cleared for timber, and to provide land for farming, early during European settlement. Very little remains. The best example is a small area (about 6 ha) known as Macquarie Hospital Bushland, recently renamed Wallumetta Forest, at the end of Twin Road, East Ryde. Here are trees of Turpentine *Syncarpia glomulifera,* Grey Ironbark *Eucalyptus paniculata,* White Stringybark *Eucalyptus globoidea* and Red Mahogany *Eucalyptus resinifera.* Grey Gum, *Eucalyptus punctata*, and smooth-barked *Angophora costata* occur near the shale/sandstone boundary. The understorey is mainly shrubby, with patches of *Pittosporum undulatum, Polyscias sambucifolia, Dodonaea triquetra, Acacia falcata* and *Acacia linifolia,* but there are also grassy open areas, some the result of disturbance, with Kangaroo Grass, *Themeda australis* and Blady Grass, *Imperata cylindrica.* Areas recently weeded are being recolonised by creeping *Commelina cyanea* and *Pratia purpurascens* and the fern *Adiantum aethiopicum.* In keeping with its intermediate character, the bushland understorey includes some species from the higher rainfall forests, such as the fern *Culcita dubia* and the shrub *Platylobium formosum,* and some from drier areas, the shrubs *Melaleuca decora* and *Acacia falcata.*

Remnants of the high rainfall Blue Gum High Forest can be seen in Darvall Park at Denistone and in Brush Farm Park at Eastwood. These are situated in parts of the municipality where the shale soils are

The wooded sandstone headlands of the northern side of Parramatta River in 1882. Putney Park today still has remnant Blackbutts, *Eucalyptus pilularis,* and *Banksia serrata* trees, as well as some of the ground species such as *Lomandra longifolia.* (Mitchell Library, State Library of New South Wales)

Here log boundaries constrain mowing and allow native understorey species to grow back beneath the Sydney Blue Gums and Turpentines on the deep clay soil of Darvall Park at Denistone. (J. Plaza, RBG, 1990)

deep. At Darvall Park, near Denistone Station, there are tall trees of the smooth-barked *Eucalyptus saligna,* along with the rough-barked *Eucalyptus pilularis, Eucalyptus paniculata, Eucalyptus resinifera, Eucalyptus acmenoides* (near its southern limit here) and *Syncarpia glomulifera.* Much of the understorey has been replaced with grass patches, but amongst the weedy shrubs and trees (*Ligustrum, Lantana, Salix, Erythrina* and *Cinnamomum*), many native understorey species are present, and their growth is being encouraged in areas protected by log barriers from mowing and trampling. Small trees of *Backhousia myrtifolia* and *Glochidion ferdinandi,* shrubs of *Bursaria spinosa, Helichrysum diosmifolium,* at least four species of *Acacia,* and moisture-loving vines including *Morinda jasminoides, Celastrus subspicatus, Cissus antarctica* and *Cissus hypoglauca* are among native plants present in these regenerating areas. At Brush Farm Park, similar trees grow on the upper slopes, but in the steep-sided, sheltered gully, fertile shale-derived soil and high rainfall support a rainforest vegetation with species not found together elsewhere in sheltered sandstone gullies or on Wianamatta Shale soils of northern Sydney. Named Brush Farm by the early settlers because of this rainforest or 'brush', its species include trees of *Cryptocarya glaucescens, Euodia micrococca, Guioa semiglauca* and *Schizomeria ovata,* shrubs of *Alectryon subcinereus* and *Eupomatia laurina,* and the climber *Aphanopetalum resinosum.* There is a very large *Trochocarpa laurina* 12 m high. The moist fertile gully soils have been particularly susceptible to weed invasion, not surprising in view of its settlement in 1794 and long farming history, before dedication as a park. In spite of its very weedy undergrowth, over 100 native species can still be found at Brush Farm Park. In most other parks on shale, for example, Stewart Park in Marsfield, and most of Denistone Park, mown grass has replaced the natural understorey.

Most of Ryde's remaining bushland is on sandstone along the Lane Cove River valley, around the edge of the municipality. Trees that occur in shale forests may grow here in gullies where soil is deep and fertile enough to support them, having been enriched by eroded clay washed down from shale areas upstream. The Lane Cove River provided a transport route for European settlers and the easily accessible flats along its banks are much modified. However, the steepness of the nearby sandstone hillsides discouraged development, and natural bushland was left, much of it later becoming the nucleus of the Lane Cove River State Recreation Area. Including most of the river's western bank, from Fairyland south of Fullers Bridge, to Browns Waterhole at the Terry Creek junction in North Epping, this is a good place to see a variety of plant communities[66 67].

Most of this State Recreation Area vegetation is open-forest, typical of sandstone hillsides and gullies. Flatter ridge-tops with woodland and heath are less common, such areas being favoured for housing, but some of their characteristic vegetation remains on upper hillslopes east of the Northern Suburbs Cemetery in Delhi Road, and near Christie Park and Blaxlands Waterfall, nort-east of Macquarie University. Two Scribbly Gums, *Eucalyptus haemastoma* and *Eucalyptus racemosa,* occur here either as woodland dominants, or emergents above shrubby heath patches, interspersed with thickets of *Kunzea ambigua; Eucalyptus racemosa* is far less abundant, as its preferred clayey soils nearer the level ridge-tops have mostly been cleared.

Between Fairyland and Fiddens Wharf open-forest on exposed hillsides predominates. Further upstream, as the valley narrows, open-forest typical of sheltered slopes becomes more common. Rocky creek banks have characteristic riparian shrubland while small stretches of mesic forest occur in the sheltered lower reaches of many creeks. These small areas are on moist, nutrient-enriched soils, and have all suffered from weed invasion.

Wider gullies and narrow riverside terraces with deeper, more fertile soil support taller trees; *Eucalyptus pilularis* and, less frequently, *Eucalyptus saligna,* up to 30 m high, together with *Syncarpia glomulifera, Eucalyptus resinifera* and *Eucalyptus paniculata.* These can be seen along the Riverside walking track north of Carter Creek. Large riverside terraces were generally cleared of timber, then used for orchards and market gardens before becoming the present-day picnic grounds. Weed invasion of the mesic native understorey has occurred in most of these nutrient-enriched sites.

Brackish sedgeswamps and mudflats appear to have been common along the river in 1788, but the 1938 weir at Fullers Bridge placed a limit on the river's tidal influence. Sediment deposited along the river, particularly over the last 100 years, has increased to provide conditions for vigorous mangrove growth, while at the same time the sedgeswamps and saltmarsh have been destroyed by draining and landfill. Although it is difficult to find a full sequence of estuarine communities together at any one place, the past pattern, delineated by slight variations in salinity and drainage, may be inferred from remnants scattered amongst various riverside sites today. *Melaleuca linariifolia* paperbarks grow along watercourses near the base of sandstone hillsides, while *Casuarina glauca* forest, 15-20 m tall, with a sparse reedy understorey of *Phragmites australis, Juncus kraussii* and *Sporobolus virginicus,* occupies low levee banks and parts of alluvial flats that are only periodically flooded. Rushlands with *Juncus kraussii, Juncus usitatus* and *Isolepis inundatus,* and reed swamps of *Phragmites australis* are found on brackish flats with waterlogged soil or standing water; today a good example can be seen behind mangroves at the mouth of Pages Creek. Saltmarsh grew on moist semi-saline low terraces behind mangrove-lined creek and river banks. A low scattered herbfield — with *Samolus repens, Cotula coronopifolia,* succulent-leaved *Sarcocornia quinqueflora,* the rush *Juncus kraussii* and sedge *Baumea juncea,* Salt Couch, *Sporobolus virginicus,* and occasional mangrove shrubs — saltmarsh is the community most affected by reclamation and landfill[67]. By contrast, the mangrove fringe along the river may

The Parramatta River, 1928, looking west towards Parramatta with Homebush Bay on the left. The Rhodes peninsula in the foreground has been industrialised but extensive mangroves and saltmarsh remain in Homebush Bay beyond, though the construction of the retaining wall proposed in 1893 (see Auburn) has altered their relative distributions. On the northern side Meadowbank, West Ryde and Ermington are still essentially agricultural with patches of forest. (Government Printing Office Collection, State Library of New South Wales)

be more extensive now than 200 years ago, as developments in the catchment have increased erosion and deposition of mud and silt, providing new opportunities for colonisation[69 70]. Low closed-forest up to 10 m tall, of Grey Mangrove, *Avicennia marina,* and the less common River Mangrove, *Aegiceras corniculatum,* forms on muddy sandbanks inundated daily by tides.

A good place to see the relationships between the different soils in Ryde is along the valley of Buffalo Creek. Beginning amongst houses on Ryde's clay soils, it flows through a sandstone gully before joining the Lane Cove River amidst mangrove-covered mudflats. In Ryde Park, near the headwaters, a few Grey Ironbarks remain of the Turpentine–Ironbark Forest that grew on the shale here. In Burrows Park the change from shale to sandstone occurs; *Eucalyptus saligna,* the Sydney Blue Gum, grows along the creek line in shale-derived clay soil, sandstone is exposed in the creek bed and the vegetation changes to open-forest, woodland and heath characteristic of sandstone, though the Narrow-leaved Apple, *Angophora bakeri,* Grey Gum, *Eucalyptus punctata,* and Blackbutt, *Eucalyptus pilularis,* indicate clay influence.

Estuarine vegetation along Buffalo Creek in 1919, with woodland on the sandstone hillsides. The area has since been filled as a tip and levelled for parkland. (Linnean Society of New South Wales)

Dense regrowth forest along Balaclava and Blaxland Roads, Eastwood, around 1912. The largest Blackbutts and Blue Gums would have been cut for timber. (Photo George Hawkins, 1912. Ryde Municipal Council Library Service, Local Studies Collection)

An important remnant of the Turpentine-Ironbark Forest is the Wallumetta Forest on Twin Road at Ryde. Bush regeneration work is removing weeds to reveal a wealth of flowering herbs and twiners growing beneath trees of Turpentine, *Syncarpia glomulifera* (right), and Grey Ironbark, *Eucalyptus paniculata* (left). (J. Plaza, RBG, 1990)

Fig. 15 Remnant of the Eastern Suburbs Banksia Scrub at Jennifer Street, La Perouse.

Fig. 16 Baumea sedges and *Viminaria juncea* shrubs (foreground), and *Melaleuca quinquenervia* paperbarks (background) grow in this coastal Freshwater Wetland at Botany.

Fig. 17 Estuarine Wetlands with extensive saltmarsh fringed by *Avicennia marina* mangroves on Towra Point.

Fig. 18 A beautiful and relatively undisturbed Freshwater Wetland, dominated by the reed *Eleocharis sphacelata*, near Lower Portland. One of the few remaining on the Nepean–Hawkesbury floodplain, it provides important habitat for waterplants and birds.

The valley deepens at the Cascades, part of the Field of Mars Reserve, and Coachwood, *Ceratopetalum apetalum,* appears, in the sheltered valley conditions. A stand of *Eucalyptus saligna* has grown up within the last 50 years on previously cleared land. Alluvial flats become broader as the creek flows past the Field of Mars Field Studies Centre and is joined by Strangers Creek. Weeds have invaded these creek lines, necessitating bush regeneration work.

Alluvial flats merge into saline wetland and although most of the saltmarsh here was destroyed in the 1950s by use as a garbage tip, traces of other wetland communities remain. Small swamps with *Phragmites* and *Typha* grow on the creek margins, backed by stands of *Casuarina glauca,* with *Melaleuca linariifolia* extending higher up creek lines. East of Pittwater Road the creek is flanked by wide bands of *Avicennia* mangroves on its route to the river. On the northern side is Sugarloaf, a sandstone knoll with relatively weed-free open-forest. *Angophora bakeri* grows here, and *Kunzea ambigua* thickets thrive on its shallow-soiled sandstone shelves. The varied vegetation along Buffalo Creek provides the outdoor resources for the Field Studies Centre, and many examples of how our lifestyles cause environmental change[74].

Cabbage Palms, *Livistona australis*, are conspicuous on the shale soils of the Narrabeen Group near Bilgola. Here are healthy adult palms in a park at Bayview. Unfortunately mowing is stopping any future regeneration. (1988)

38 Warringah

Warringah is a large shire with a very varied range of habitats. Its proximity to the Pacific Ocean and Pittwater, the high rainfall, between 1,200 and 1,400 mm per year, and absence of frosts influence much of its vegetation.

On 22nd August 1788 Governor Phillip landed in Manly Cove to explore the coastline as far as Broken Bay. During the day they were held up for two hours by the tide at Narrabeen Lagoon and camped, according to John White 'on a little eminence by the side of a cabbage tree swamp, about half a mile [800 m] from the run of the tide. Here the whole party got as much cabbage, to eat with their salt provisions, as they chose'[23]. The 'cabbage tree' was of course the Cabbage Palm, *Livistona australis,* which is still conspicuous along parts of Deep and Middle Creeks and around Bayview, Bilgola and Avalon. The next day, around midday, Phillip's party reached Pittwater, 'but finding the country round this part very rugged, and the distance too great for our stock of provisions, we returned to the sea shore, in order to examine the south part of the entrance into the Broken Bay. This, like every other part of the country we have seen, had a very indifferent aspect,' wrote White, disappointed at the lack of open country suitable for farming. However he does note that: 'From the entrance of Port Jackson to Broken Bay, in some places from fifty to a hundred, in others to two hundred yards [about 200 m] distance from the sea, the coast indeed is very pleasant, and tolerably clear of wood; the earth a kind of adhesive clay, covered with a thick and short sour grass'. This appears to be a reference to the headlands from Long Reef northwards, where outcropping Narrabeen Group shales weather to clay soils which appear to have had grasslands of Kangaroo Grass, *Themeda australis.* Such grasslands are still found on the New South Wales central coast south of Catherine Hill Bay near Swansea.

On the return journey the party boiled and ate some beans from a coastal vine, *Canavalia maritima.* The governor and White were 'seized with violent vomiting. We drank warm water, which, carrying the load freely from our stomachs, gave us immediate relief.' White was impressed by the richness of the local flora. 'During our return, we picked up, in the distance of about half a mile, twenty-five flowers of plants and shrubs of different genera and species.' The August flowering display of coastal heath plants is just as impressive today, and *Canavalia* still grows at Palm Beach and Long Reef.

Most of Warringah's natural vegetation is found on the Hawkesbury Sandstone landscapes, extensive tracts of which are included in the major conservation reserves, Davidson Park State Recreation Area, Manly-Warringah War Memorial Park and of course Ku-ring-gai Chase National Park, as well as in smaller reserves such as Katandra Bushland Sanctuary (Mona Vale), Dee Why Head and Hudson Park at Avalon. Vegetation here follows similar patterns to that elsewhere, ranging in structure from open-forest on sheltered sites, lower hillslopes and along creek lines, to woodland on ridges. Exposed aspects, particularly west-facing, or exposed coastal headlands will have heath or low woodland vegetation, while sheltered south and south-east aspects will have taller forest. Localised changes in soil and drainage, often related to interbedded shale lenses in the sandstone, may result in permanently wet soaks with sedgeland or wet heath. Similarly deeper red gravelly soils on some of the broader ridges may have forest vegetation.

Typical examples of the sandstone vegetation can be seen easily from Wakehurst Parkway in Manly-Warringah War Memorial Park or around Bantry Bay in Davidson Park State Recreation Area, particularly from the Frenchs Forest Lookout at the northern end. Here on shallow soil on the rocky exposed ridges, heath and scrub predominate. Small patches of heath, with shrubs generally less than 1 m high, are found on exposed rocky outcrops, particularly on the Allambie Heights side of the War Memorial Park. Common species are *Leptospermum attenuatum, Darwinia fascicularis* and *Kunzea ambigua.* On more sheltered sites, with perhaps deeper soil, the vegetation is taller, grading into closed-scrub up to 4 m high, dominated by *Allocasuarina distyla, Banksia ericifolia* and *Hakea teretifolia.* Open-scrub with similar species but including *Banksia oblongifolia* is characteristic of wetter sites, while *Angophora hispida* is more common where soils are drier. Small localised patches of Port Jackson Mallee, *Eucalyptus obtusiflora,* occur in heath near Allambie Heights while another mallee, Yellow-top Ash, *Eucalyptus luehmanniana,* is found in similar areas in Bantry Bay but not in War Memorial Park. *Eucalyptus luehmanniana* is also found on south-facing slopes below Frenchs Forest Lookout and in similar aspects in Davidson Park.

Drainage conditions play an important role in the distribution of many species. Closed-scrub may indicate periodically poorly drained soils, while in permanently poorly drained or swampy sites are patches of sedge and shrub-swamp. Common species in these conditions are *Baeckea imbricata, Epacris pulchella, Leptospermum squarrosum* and *Hakea teretifolia.* Flowering Christmas Bells, *Blandfordia nobilis,* may be conspicuous here after a fire. They are particularly prolific in poorly-drained areas of Frenchs Forest Cemetery, and north of Cascades.

On the western part of the War Memorial Park, below the Wakehurst Parkway, the edge of the scrub is marked by rocky cliff-lines up to 6 m high, with overhangs and caves. Moist rock ledges here provide an excellent habitat for ferns, *Culcita dubia, Gleichenia dicarpa* and *Gleichenia rupestris,* and the small shrub, *Epacris crassifolia.* An unusual orchid, *Rimicola elliptica,* has been recorded on these moist rock ledges, and the primitive fernlike *Psilotum nudum* is found on dry rocky cliff-lines.

Low open-woodland occurs on the lower sides of ridges and shallow gullies, and is widespread in War Memorial Park and Bantry Bay. It is dominated by small trees generally less than 10 m in height (commonly 5–8 m) and forms an open, scattered canopy, though density varies. Common trees are Red Bloodwood, *Eucalyptus gummifera,* Scribbly Gum, *Eucalyptus haemastoma, Stringybarks, Eucalyptus oblonga* and *Eucalyptus capitellata,* Sydney Peppermint, *Eucalyptus piperita,* Black Ash, *Eucalyptus sieberi* and Narrow-leaved Apple, *Angophora bakeri.* Understorey shrubs are about 2 m high. There are some scrub community species, such as *Banksia ericifolia* and *Allocasuarina distyla,* as well as additional species such as the tall *Banksia serrata* and the smaller shrubs, *Pultenaea stipularis, Boronia ledifolia* and *Woollsia pungens.*

In sheltered gullies, trees are taller (over 10 m) and form woodland or, where denser, open-forest vegetation. This is of limited extent in the War Memorial Park but more extensive in Davidson Park and Bantry Bay. The main trees are Sydney Peppermint, *Eucalyptus piperita,* and smooth-barked *Angophora costata.* The understorey is generally shrubby with species similar to low woodland, together with others such as *Dodonaea triquetra* and *Acacia terminalis,* which evidently prefer the more sheltered conditions. Variations in species composition may be caused by outcropping shale strata that weather to clayey soil and alter drainage conditions. Species on one such clayey area included *Acacia myrtifolia, Boronia pinnata, Grevillea linearifolia, Dodonaea triquetra, Pultenaea daphnoides* and *Pultenaea hispidula,* the latter an uncommon species in the Sydney area. Some species are restricted to creek lines; in War Memorial Park, for instance, the shrubs *Callicoma serratifolia, Baeckea linifolia* and *Austromyrtus tenuifolia* and the attractive sedge *Restio tetraphyllus* may be found. Along Middle Harbour Creek are *Lomatia myricoides, Stenocarpus salignus, Backhousia myrtifolia, Leptospermum grandifolium* and *Ceratopetalum apetalum.*

Similar types of vegetation are found in the smaller sandstone reserves of Warringah although there may be local differences in species present. For example, there is a population of the rare *Darwinia procera* in Davidson Park near the Cascades, while Katandra Bushland Sanctuary has small sheltered rainforest gullies, *Boronia thujona* and the locally uncommon *Boronia mollis.*

One particular type of vegetation not in War Memorial Park or Bantry Bay, and now very rare, is found on the remnants of 'lateritic' or ironstone soils of broad ridge-tops at Frenchs Forest, Belrose, Terrey Hills and Duffys Forest. There is some discussion about the origin and nature of these soils but the vegetation is quite characteristic. Most of it has been destroyed during quarrying of the gravels for road materials, or to make way for housing, but there is a remnant at the eastern corner of the junction of Mona Vale Road and Forest Way, Terrey Hills. Here is woodland 10-12 m high with Black Ashes, *Eucalyptus sieberi,* Brown Stringybarks, *Eucalyptus capitellata,* smooth-barked *Angophora costata,* Red Bloodwoods, *Eucalyptus gummifera,* and occasional Scribbly Gums, *Eucalyptus haemastoma.* The understorey is shrubby and ranges from 1 to 4 m in height depending generally on the time since the last fire. Common shrub species are *Acacia myrtifolia, Banksia spinulosa, Ceratopetalum gummiferum, Bossiaea obcordata, Micrantheum ericoides* and *Grevillea linearifolia.* The species composition is different from woodland on the surrounding Hawkesbury Sandstone. One feature is the high number of species of Proteaceae present, 19 species being recorded here. This represents about half of the species of Proteaceae recorded in Ku-ring-gai Chase National Park. Several small populations of the rare shrub *Grevillea caleyi,* a locally endemic species, occur here. This species used to be found at Frenchs Forest and Belrose but has been destroyed by a shopping complex and housing developments. Only small populations have been protected in Ku-ring-gai Chase National Park and it is important that further populations be conserved before they are all destroyed.

There is not space to describe every area of sandstone vegetation in the Shire individually, though mention should be made of the catchment

A party of shingle-splitters on the shores of Middle Harbour about 1880 continue a trade begun in early colonial days. In 1813, for example, Governor Macquarie instructed each three-man gang to supply 4,000 shingles per week. Cut from *Casuarina* and *Allocasuarina* trees, known locally as she-oaks, shingles roofed many of Sydney's houses in the nineteenth century. (Mitchell Library, State Library of New South Wales)

of Deep Creek. This is a large area of virtually undisturbed vegetation bounded by Mona Vale Road and Forest Way and draining to Narrabeen Lakes. In particular, the vegetation on the ridges and slopes is in excellent condition, unaffected by exotic weed invasion and little influenced by human activity. There are particularly extensive areas of heath here, very little explored botanically, and apparently very rich in species, including some rather odd occurrences. For example, in a recent survey we found localised populations of *Platysace stephensonii* and *Allocasuarina nana*. The first is a rare species now known from a few rocky outcrops in Ku-ring-gai Chase, though it was recorded on the sandy lowlands behind Rose Bay (see Woollahra) in the 1890s, and has recently also been found near Botany Bay. The second species, *Allocasuarina nana,* is a major component of heath in the upper Blue Mountains at elevations over 800 m. The very localised and isolated group of about 30 plants on a ridge at Deep Creek may be a relict population perhaps indicating the occurrence of 'Blue Mountains' type heath at Deep Creek during the last glacial period 12-20,000 years ago. The vegetation of Deep Creek needs to be investigated, and in particular protected from piecemeal development of the catchment margins that will cause long-term deterioration, particularly in the gully flora.

About 1,000 native plant species are found in Warringah. These occupy a variety of environments — in addition to the Hawkesbury Sandstone landscapes there are coastal beaches, cliffs and lagoons, Narrabeen Group sandstones and shales, shale-covered plateaus, and alluvial stretches along creeks.

Along the Barrenjoey Peninsula, north of Long Reef, shales of the Narrabeen Formation outcrop below the Hawkesbury Sandstone forming a fine clay soil. Here on the lower hillslopes, along the ocean shoreline and around the shores of Pittwater, is a characteristic Spotted Gum Forest, particularly evident around Clareville, Newport and Avalon. The Spotted Gum, *Eucalyptus maculata*, is the most conspicuous tree, with some Grey Ironbark, *Eucalyptus paniculata,* Turpentine, *Syncarpia glomulifera,* and Forest Oak, *Allocasuarina torulosa.* Where the site is dry and exposed, there are trees of the related *Allocasuarina littoralis,* shrubs of *Pultenaea flexilis, Hakea sericea,* and the smaller *Acrotriche divaricata* and *Oxylobium ilicifolium* in the understorey. In moister, more sheltered situations, the place of shrubs is taken by small 'rainforest' trees including *Glochidion ferdinandi, Notelaea longifolia, Synoum glandulosum, Livistona australis,* mesic herbs, ferns and vines.

Ocean beaches and lagoon systems alternating with bluff headlands and rocky escarpments are a particularly characteristic element of the Warringah landscape. Unfortunately most of the native vegetation near the beaches has been destroyed by recreation and residential use, though remnants occur here and there to give us an idea of its former appearance. The degree of loss, however, gives some idea what might happen to our less disturbed beaches away from Sydney, unless we take more care to protect their natural values. Sand-binding grasses, *Spinifex hirsutus* and *Festuca littoralis,* would have occupied the foredune with *Lomandra longifolia* clumps on the dune crests. Many of these areas have been replanted with the introduced Marram Grass, *Ammophila arenaria,* for stabilisation. On sheltered sites behind the dunes was low scrub with *Leptospermum laevigatum, Banksia integrifolia, Acacia longifolia* var. *sophorae, Acmena smithii* and the scramblers, *Stephania japonica, Kennedia rubicunda* and *Commelina cyanea.* Introduced species now common are *Coprosma repens* shrubs and the creeper *Hydrocotyle bonariensis.*

Headlands with clay soil of the Narrabeen Group geology in northern Warringah, at Narrabeen, Turimetta, Mona Vale and Long Reef, appear to have once had grassland of the native Kangaroo Grass, *Themeda australis,* but were probably used for grazing last century. By 1918 the botanist Arthur Hamilton was reporting: 'Throughout this area in which the soil is comparatively rich, the indigenous vegetation has been removed, and the ground dedicated to a pasture of "Couch-grass" *Cynodon dactylon* Rich., which creeps out to the verge of the ocean headland or the dune embankment'[42]. Other native species still found in remaining patches of grassland include *Patersonia longifolia, Thysanotus tuberosus, Burchardia umbellata* and *Lasiopetalum ferrugineum.* Warringah Council is currently developing management regimes using fire to replace mowing to encourage natural regrowth.

South of Dee Why Beach the headlands are of Hawkesbury Sandstone and had coastal heath, no good for grazing stock. There are good examples remaining at Dee Why Head and, of course, at North Head in Manly. At Dee Why Head common species are *Banksia ericifolia, Baeckea imbricata* and *Lomandra longifolia.* Hamilton records *Melaleuca nodosa* as abundant in 1918 and describes the vegetation on the exposed southern slope as 'low and closely packed, creeping over, or clinging to the verges of the boulders and benches, and rooting in the soil pockets among the rock masses'. It still does.

This photo of Dee Why Lagoon in the early 1900s shows a tranquil scene of Swamp Oaks, *Casuarina glauca*, and sedges, probably here *Baumea articulata*. (Government Printing Office Collection, State Library of New South Wales)

In 1925 Narrabeen was a holiday resort, still to feel the impact of the suburbs. Housing concentrated along the beachfront, forests fringed the undisturbed lagoon, and the sandstone plateau of Collaroy still had its characteristic shrubby woodland. (Government Printing Office Collection, State Library of New South Wales)

Coastal heath on sandstone at Curl Curl on the Coastal Walkway is in good condition. (1988)

The coastal lagoons and swamps have been variously altered. At Curl Curl or Harbord Lagoon almost all natural vegetation has been removed except for a reedy fringe of *Phragmites australis*. Further north, Dee Why Lagoon has heath and scrub on low sand ridges on the southern side of the lagoon. Narrabeen Lagoon has fringes of *Casuarina glauca* woodland and *Phragmites*, while on the alluvial fan of Deep Creek are remnants of the original Swamp Mahogany, *Eucalyptus robusta*, forest. Prior to settlement *Eucalyptus robusta* probably occurred throughout the low-lying lands of Dee Why, Curl Curl, Warriewood, Narrabeen, Mona Vale, Bayview, Newport and Avalon. The best areas of Swamp Mahogany, however, are in the Warriewood Wetlands on lower Mullet Creek. This is the last sizeable stand of this vegetation near Sydney and it is important that it is retained and restored.

The predominance of poor, sandy soils on the sandstone landscapes and the low-lying poorly-drained estuarine flats meant that the Warringah area held little potential to the early colonists who were seeking agricultural lands. There was little development in the area in the nineteenth century, apart from market gardens and poultry farms in places such as Mona Vale and Oxford Falls. These took advantage of the mild climate and absence of frosts along the coast, and in places soil enriched by volcanic dykes.

In the twentieth century and particularly after World War I, the coastal beaches became popular holiday resorts. Small cottages were built behind the beaches from Manly to Newport, as motor cars made these areas increasingly accessible.

Intensive suburban housing began to spread only after World War II, first along the coastal beaches and then up on to the sandstone plateaus. The amount of residential development since then has been enormous, and very extensive areas of previously undisturbed bushland have been destroyed. Early suburban residents who built their houses in bushland have seen it disappear around them. Typical are the reminiscences of Helen Veitch of Frenchs Forest, describing changes since 1940:

> People came from Chatswood to picnic along the old road to Pymble (now Forest Way). Areas of laterite still had Waratahs, *Xylomelum* and *Banksia, Eucalyptus sieberi*, Bloodwoods, Stringybarks and three species of *Angophora. Lobelia* and *Blandfordia* were common on Frenchs Forest Road; children on their way to school carried bunches of blue orchids. *Scaevola, Epacris, Patersonia* grew prolifically along deep unmade gutters. Hardly a vestige of forest left after 48 years of "progress"; very few blackbutts remain — maybe a dozen in Blackbutts Road. The undergrowth — five different species of *Persoonia* — has almost disappeared.[75]

Where bushland has been retained less obvious changes are still occurring as increases in run-off and pollution allow exotic weeds to invade bushland gullies, and altered fire frequencies result in changes in understorey shrub cover. Trees such as the distinctive Spotted Gums along Pittwater, though often carefully retained during housing construction, are gradually ageing and dying, to be replaced by non-local species. And so the particular local and often unique character of each area is lost — to be replaced by a colourful but ubiquitous array of Hibiscus, Azaleas and Willow Gums.

39 Willoughby

The gently undulating plateau occupied by the houses of Artarmon, Willoughby and Chatswood, the wooded Middle Harbour hillsides between Castlecove and Northbridge, and the Lane Cove River as it winds between Fullers Bridge and Epping Road are three distinctive Willoughby landscapes; dependent on the different geologies of shale, sandstone and tidal flats respectively. Each once had characteristic vegetation.

About half of Willoughby, its higher ridges and main plateau, is covered by a Wianamatta Shale mantle that weathers to clay-rich, relatively fertile soil. Chatswood, most of the suburb Willoughby, and the higher parts of Artarmon and Naremburn once had magnificent Blue Gum High Forest, with its tall trees of Sydney Blue Gum, *Eucalyptus saligna*, and Blackbutt, *Eucalyptus pilularis*. Cut and burnt by the early timber getters and settlers, none of this forest survives in the municipality today, though where sandstone gullies have been enriched by downwashed clay, some of the High Forest trees and understorey plants can still be seen. In Artarmon Reserve, for example, beside the creek, are a few Blue Gums, together with Blackbutts, and trees of *Syncarpia glomulifera, Angophora costata, Eucalyptus piperita*, and *Eucalyptus resinifera*. Despite disturbance from the sewer line along the creek, and nutrient enrichment from suburban areas upstream that have encouraged weed growth, mesic native understorey species here include small trees of *Notelaea longifolia* and *Polyscias sambucifolia*, the shrubs *Breynia oblongifolia, Leucopogon juniperinus* and *Pultenaea flexilis*, and vines *Clematis* and *Tylophora barbata*. The creek line is terminated abruptly by the oval, and vegetation elsewhere in the reserve is of typical sandstone flora.

Farming developed in central Willoughby in the mid-nineteenth century, after the Blue Gum High Forest had been logged. Suburban development on the relatively level shale country at Chatswood and Artarmon followed the opening of the North Shore railway in 1890 and continued after World War I. The Northbridge suspension bridge, also completed in 1890, opened further areas for suburbs, but the steeper sandstone country was difficult to build on and bushland areas associated with the major waterways were not developed until after World War II. An important exception, Castlecrag, was used by architect Walter Burley Griffin, designer of Canberra, for developing his ideas on integrating suburban housing with the natural bushland. As a result Castlecrag has quite a number of small bushland pockets set within the suburban matrix, many of which are deteriorating through exotic weed invasion.

The Hawkesbury Sandstone underlying the shale is exposed on hillsides and ridges along both Middle Harbour and the Lane Cove River, and parks and reserves with bushland conserve examples of its characteristic vegetation[76]. In the Harold Reid Reserve at Middlecove, and A.C. Press Reserve at Castlecove (linked by the North Arm Walking Track), steep wooded slopes rise almost 100 m above the water of Middle Harbour. Woodland of the Scribbly Gums, *Eucalyptus haemastoma* and *Eucalyptus racemosa,* Red Bloodwood, *Eucalyptus gummifera,* Black She-Oak, *Allocasuarina littoralis* and Old Man Banksia, *Banksia serrata,* and a variety of flowering shrubs, herbs and grasses may be found on the ridge-tops.

The Sugarloaf, the sculptured sandstone crest of the Harold Reid Reserve, emerges from amongst a host of shrubs. In autumn colourful flowers include *Acacia linifolia, Banksia ericifolia, Banksia marginata, Crowea saligna, Epacris longiflora* and *Phyllota phylicoides.* Conspicuous amongst the monocotyledons are metre-tall loose clumps of *Caustis pentandra,* trunkless *Xanthorrhoea* Grass Trees and stiff-leaved *Lomandra longifolia.* Local changes in species composition are generally linked to differences in soil and drainage. For example, where the poor sandy soil has some clay enrichment, *Eucalyptus racemosa* may replace the more common *Eucalyptus haemastoma.* Grey Gums, *Eucalyptus punctata,* also indicate local clay influence, and certainly occurred closer to the shale boundary; a few can be seen today on the edge of the H.D. Robb Reserve. Trees of *Eucalyptus sieberi,* the Silver-top Ash, are restricted to just below the summit of

This Blue Gum, *Eucalyptus saligna*, stood 'in a Chinaman's garden' in Chatswood, near Boundary Road, in 1885. With a height of 84 feet (25 m) and a girth of 34 feet (10 m), this sadly decaying giant provides evidence of the size of trees in Willoughby's Blue Gum High Forest.

Scribbly Gums, *Eucalyptus haemastoma*, overlook Sugarloaf Bay from Harold Reid Reserve, the best natural area on the western side of Middle Harbour. (1988)

Suspension Bridge, Northbridge, across the forested valley of Flat Rock Creek about the turn of the century. (Macleay Museum)

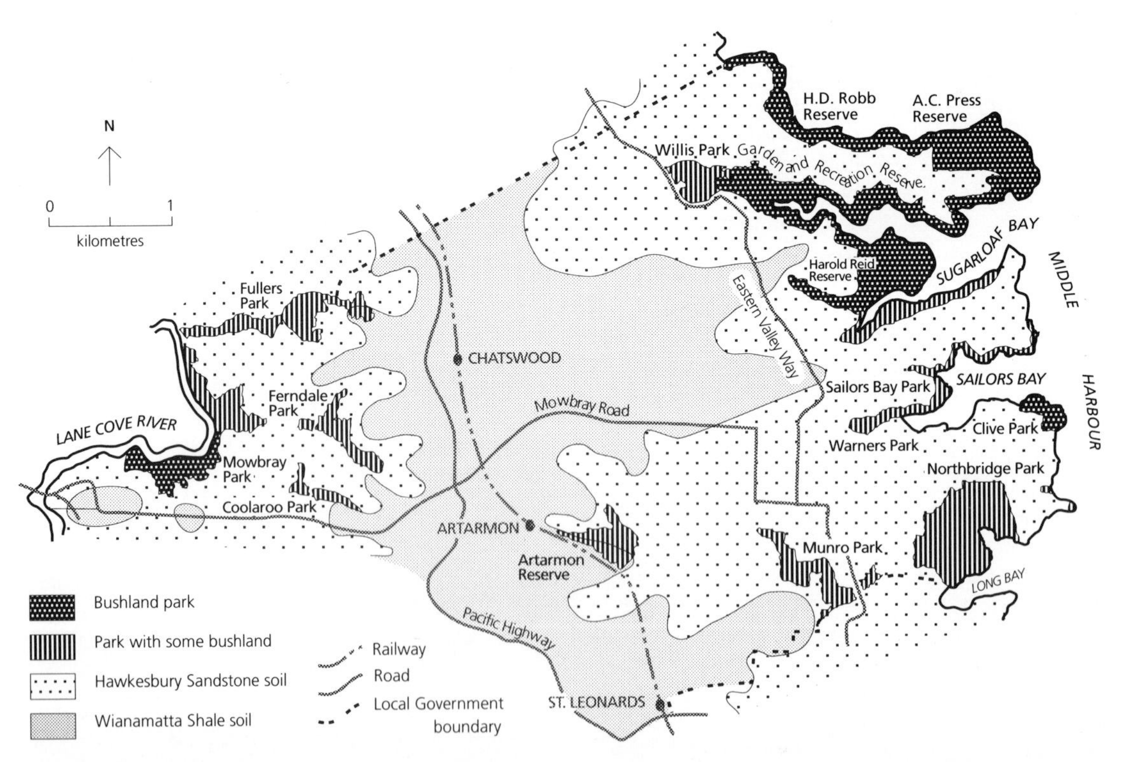

Parks and reserves with bushland areas in Willoughby Municipality. The extent of Wianamatta Shale and Hawkesbury Sandstone is shown.

the Harold Reid Reserve on the more exposed north-facing side.

On the heights of the A. C. Press Reserve, heath areas are interspersed with the woodland. *Angophora hispida, Hakea teretifolia,* and *Banksia ericifolia* are shrubs characteristic of these, often wetter, sites. Vegetation on this former Commonwealth land is in good condition, and quite scarce around Middle Harbour, being on the type of flatter ridge-top land that has generally been used for housing. On the slopes are some quite large trees of *Angophora costata* and *Eucalyptus gummifera,* together with Sydney Peppermint, *Eucalyptus piperita.* Small Blueberry Ash trees, *Elaeocarpus reticulatus,* and Christmas Bush, *Ceratopetalum gummiferum,* are common, also in places *Banksia serrata, Allocasuarina littoralis* and *Allocasuarina torulosa.* On more protected or fertile slopes the shrubs include *Grevillea linearifolia, Dodonaea triquetra* and *Pomaderris.* The creeping *Leucopogon amplexicaulis* grows on moist rock ledges or creek margins on the south-facing lower slopes.

Along Willoughby's sandstone-lined creeks, the vigorous ferns *Culcita dubia* and *Gleichenia dicarpa* replace the shrubs. Deeper, more sheltered gullies may contain Coachwood trees, *Ceratopetalum apetalum,* and Tree Ferns, *Cyathea*; some survive in Flat Rock Gully, for example. Landfill operations upstream here have already destroyed large areas of bush over the last 20 years and should be constrained, and filled areas rehabilitated to reduce further degradation of the remaining bushland. Small pockets of creekside vegetation may also be found along Bay, Camp, and Sugarloaf Creeks, in Retreat and Castlehaven Reserves, and in Munro and Clive Parks. The falls on Sugarloaf Creek are worth seeing after rain; recent bush regeneration work here has removed weedy overgrowth, restoring easy access.

Where major creeks meet the harbour, the tidal mud and sandflats are colonised by mangroves and *Casuarina.* There are healthy stands of the Grey Mangrove, *Avicennia marina,* along the north and south arms of Sugarloaf Bay. Samphire plants, *Sarcocornia quinqueflora,* may occur with the mangroves, fringed by narrow bands of *Casuarina glauca,* forming Swamp Oak forest. *Phragmites, Juncus* and the paperbark, *Melaleuca styphelioides,* occur sporadically amongst the *Casuarina.*

On the western side of Willoughby, the vegetation patterns along the Lane Cove River are similar to those on Middle Harbour, but with some local differences due to the gentler topography and larger catchments. Mudbanks and riverflats are larger. Vegetation in Mowbray Park at Chatswood West is an interesting complex of plant communities[77]. In addition to the dominant *Avicennia* mangroves, there are shrubs of *Aegiceras corniculatum,* the River Mangrove, along the water's edge. On river flats behind the mangroves are areas of reeds, *Phragmites australis,* patches of salt-tolerant herbs including *Sarcocornia quinqueflora* and *Samolus repens,* and sizeable stands of Swamp Oak fringe forest. On a deep sandy terrace there is woodland of Blackbutt and Grey Ironbark, *Eucalyptus paniculata,* with a number of small tree species including *Xylomelum pyriforme,* the Woody Pear. *Melaleuca linariifolia* is the paperbark that grows near the base of the rocky hillside. Higher up, the exposed sandstone slopes have open-forest with impressive Blackbutts amongst the *Eucalyptus piperita* and *Angophora costata,* and a diverse understorey. In more sheltered situations, large Port Jackson Figs, *Ficus rubiginosa,* grow with small trees of *Clerodendrum tomentosum, Notelaea longifolia, Rapanea variabilis, Glochidion ferdinandi, Pittosporum undulatum* and Bird's Nest Fern, *Asplenium australasicum.* Along the creek grow a variety of ground ferns, *Cyathea cooperi* Tree Fern, the climber *Smilax glyciphylla,* and small Black Wattle trees, *Callicoma serratifolia.* Although describing only part of the plant richness that can be found in Mowbray Park, this list emphasises the benefits, even within a small bushland park, of conserving as wide a range of habitats as possible.

The Southern Suburbs

Sydney's southern suburbs are included within the Shire of Sutherland which occupies all of the land south of the Georges River and Botany Bay. Only the northern part of this large shire is described, that is, north of Port Hacking. The extensive Royal and Heathcote National Parks to the south are beyond the scope of this book.

40 Sutherland

The first Sydney plants known to science were those collected at Kurnell by Joseph Banks and Dr Daniel Solander who landed from Captain Cook's *Endeavour* in April 1770. Their landing place, now part of Botany Bay National Park but farmed last century, has stretches of mown grass, planted with trees from other parts of Australia. A few naturally occurring Bangalays, *Eucalyptus botryoides,* survive, remnants of the original woodland, and small groves of the Swamp Oak, *Casuarina glauca,* and paperbark, *Melaleuca quinquenervia,* still grow along the creeks, as they would have done in 1770. Beside one that provided Cook's party with fresh water, are some *Backhousia myrtifolia* and *Syzygium oleosum* trees and Cabbage Palms, *Livistona australis.* Some of the ferns carpeting the creek banks, *Culcita dubia, Adiantum aethiopicum* and the tree-fern-like *Todea barbara,* are also descendants of the pre-European vegetation.

Further along the shore, at Inscription Point, are dense, windswept thickets of the fine-leaved *Melaleuca armillaris,* just as when Banks and Solander botanised here. Their original specimens of *Westringia fruticosa* and *Kennedia rubicunda,* now in the National Herbarium at Sydney's Royal Botanic Gardens, could have been collected in this heath. Across Solander Drive are sand dunes with low forest and scrub. From here could have come their specimens of *Banksia serrata, Banksia integrifolia, Acacia longifolia, Hibbertia scandens, Persoonia levis, Xylomelum pyriforme* and *Angophora costata.* In April, the cream flowers of *Acacia suaveolens* and *Acacia ulicifolia,* the buds of *Acacia terminalis,* and bright yellow-brown pea flowers of *Bossiaea heterophylla* would have caught their attention. They may have noticed the green-flowered *Grevillea mucronulata,* and the herb *Pomax umbellata.* From dune crests near today's Kurnell Lookout, they would have had a panoramic view of sand dunes and swamps, with the mangroves of Quibray Bay in the distance. No wonder the young Joseph Banks spent so much of his time 'in the woods, botanizing as usual'[78].

Cook, Banks and Solander had landed on the north-eastern end of the broad peninsula separating the Georges River and Botany Bay from Port Hacking. Today Sutherland Shire includes all this land westward as far as Menai, the southern suburbs along the Princes Highway from Loftus to Waterfall, and Bundeena on the southern shore of Port Hacking. The geology of the shire is predominantly Hawkesbury Sandstone, covered in places with shale or sand. These three geological units — sandstone, shale, and sand — determine the soils and the main groupings of vegetation. Narrabeen Group shales outcrop towards the south of Royal National Park, outside the scope of this chapter.

Sutherland's main suburban area occupies the broad shale-capped ridge between Loftus and Cronulla. Prior to European settlement, the clay soils here would have been covered with Turpentine-Ironbark Forest similar to that of the inner western suburbs. In Sutherland shire however, Blackbutt, *Eucalyptus pilularis,* and Grey Ironbark, *Eucalyptus paniculata,* were the dominant trees, though Turpentine, *Syncarpia glomulifera,* Red Mahogany, *Eucalyptus resinifera,* and White Stringybark, *Eucalyptus globoidea,* were also present. For example, near Cronulla in the 1860s grew 'Iron Bark, Grey and White Gum, Mahogany, Stringybark, and Oak'[79]. Shrubs and small trees on the shale areas included *Polyscias sambucifolia, Dodonaea triquetra, Acacia parramattensis, Glochidion ferdinandi, Breynia oblongifolia, Bursaria spinosa* and *Pittosporum undulatum.* The forest followed the east-west shale spine from Sutherland to Woolooware, along President Avenue and The Kingsway, and the higher parts of the ridges followed by the Princes Highway, North West Arm Road and Burraneer Bay Road. At the old quarry in Kirrawee, walls of shale 8–10 m high give a good idea of its thickness. Shale capping extends along the Princes Highway as far as Loftus, and recurs on the high parts of Engadine and Heathcote — almost all has been cleared. At Menai, along Old Illawarra Road, remnants of the original shale vegetation survive at Thorpes Forest. Turpentines in abundance, together with Grey Ironbark, White Stringybark, and Forest

Although modified by this time, vegetation at Cronulla Beach in 1905 showed characteristic zonation. Clumps of *Festuca* grass grow in sand facing the sea, *Acacia longifolia* var. *sophorae* covers the lee side of the foredune, and small *Banksia integrifolia* trees grow further back. (Sutherland Shire Libraries and Information Service, Local Studies Collection)

Tussocks of *Juncus kraussii* and scattered mangroves, *Avicennia marina,* persisted in Gwawley Bay despite Holt's extensive oyster canals. However, these natural remnants were to disappear completely under the suburban canal subdivision of Sylvania Waters. (Sutherland Shire Libraries and Information Service, Local Studies Collection)

Red Gum, *Eucalyptus tereticornis,* grow here. The Forest Red Gums suggest affinities with the drier woodlands of the Cumberland Plain further west. The early development of Sutherland's outlying suburbs generally followed the shale, which provided the only agricultural soils in an otherwise harsh landscape. The small shale remnants at Menai and in Royal National Park at Loftus have special botanical value.

Thinner shale near the transition to underlying sandstone weathers to lower nutrient soils. Open-forest of Scribbly Gums, *Eucalyptus racemosa* and *Eucalyptus racemosa/E. haemastoma* intergrades, together with Blackbutt, Grey Ironbark and a few Turpentines, would have been widespread.

Away from the shale influence, the open-forests, woodlands and heaths characteristic of nutrient-poor, sandy soils of the Hawkesbury Sandstone landforms extended for miles. These would have had woodlands with *Eucalyptus haemastoma,* Scribbly Gums, small *Banksia serrata* trees, and the Red Bloodwood, *Eucalyptus gummifera,* with an understorey including giant pink flower heads of the Gymea Lily, *Doryanthes excelsa,* and a diversity of shrubs — *Leptospermum attenuatum, Ricinocarpos pinifolius,* species of *Grevillea, Hakea, Boronia, Eriostemon, Acacia* and *Dillwynia* among them — together with grasses, sedges, and some ferns. Heath replaced the woodland where drainage was impeded by shale layers or depressions in the sandstone. Shrub species common in heath included *Banksia ericifolia,* often in dense thickets 3-4 m tall, *Angophora hispida,* flowering with a lavish display of large cream eucalypt-like flowers, and the prickly *Hakea teretifolia.* On exposed hillsides, particularly upper slopes where deep soils have not had a chance to accumulate, vegetation characteristic of the ridge-tops continued. Today such vegetation is found on western parts and north-facing slopes of Kareela, Oyster Bay, Como and Bonnet Bay and remnants persist in Bates Drive, Kareela; opposite Salisbury Golf Course; on ridge edges at Loftus and Bonnet Bay; and in Burraneer Park. On more sheltered south- and east-facing slopes, along drainage lines, and where deeper soils have accumulated, open-forest replaced woodland. Along south-facing slopes in the waterfront suburbs from

In 1770 the wind-swept heath of Endeavour Heights looked much the same as today; Joseph Banks could have collected his specimens of *Baeckea imbricata* and *Correa alba* here. (1990)

Grays Point to Cronulla, water views are still framed by Sydney Peppermints, *Eucalyptus piperita,* and the pink trunks and twisted branches of the smooth-barked *Angophora costata,* characteristic of this topography. There are also Blackbutt, *Eucalyptus pilularis,* Black She-oak, *Allocasuarina littoralis,* Christmas Bush, *Ceratopetalum gummiferum,* and Blueberry Ash, *Elaeocarpus reticulatus.* Similar communities can be seen along creeks draining into Bonnet Bay and Oyster Bay.

Along the Georges River and Port Hacking, open-forest with Bangalay, *Eucalyptus botryoides,* and *Elaeocarpus reticulatus, Cupaniopsis anacardioides,* and Forest Red Gum, *Eucalyptus tereticornis* would have been found, along with small trees of *Pittosporum undulatum, Glochidion ferdinandi, Banksia integrifolia* and a few figs, *Ficus rubiginosa.* The more mesic of these species, together with others including *Acmena smithii, Syzygium oleosum* and several vines, formed patches of littoral rainforest such as near Bundeena and Towra Point.

Magnificent areas of sandstone vegetation remain in the Royal National Park south of Sutherland, where there is a diversity of vegetation ranging from heath to rainforest. We give much credit to the foresight of those who established Australia's first national park here in 1879 (the second national park in the world) as we see expansion of housing onto similar country further west, and how hard it is, 100 years later, to protect such areas for our future long-term enjoyment in the face of short-term ambition.

Since the 1950s, suburbs have proliferated, spreading away from the early agricultural settlements along the shale-capped ridges, on to the undisturbed woodlands and heaths of the sandstone. Ridge-tops and any level sites were built on, leaving the main suburban axis from Sutherland to Cronulla almost devoid of publicly owned bushland. Although the surrounding national parks give the impression of abundant bush, it is almost impossible to find any in parks or reserves amongst the older suburbs. As early as 1924, for example, Frank Cridland described:

> a beautiful reserve known as Darook Park. Its existence is unknown to the majority of permanent Cronulla residents. Hidden away in a protected gully running down from Nicholson Parade to Gunnamatta Bay, it is a perfect gem of a park, left in its natural state, except for a winding track cut through the undergrowth to the picturesque sand-dunes and grassy banks on the water-front. The vegetation, owing to the sheltered position and rich damp soil, is dense and semi-tropical, and a complete contrast to that of the surrounding country. The native grape, wild jasmine, and Wonga vine grow there. Creepers and parasitic vines batten on the gum-trees.

A footnote reveals its fate: 'Since the above was written Darook Park has been "improved" by the removal of the undergrowth. About the same time a fierce fire ravaged the gully. It is doubtful whether the park will ever recover from the two disasters and regain its former natural beauty'[80].

Over half a century later, in supposedly enlightened times, it is all too common to find blinkered and insensitive planning repeating the same mistakes. Recent subdivisions, like those at Menai, leave inadequate amounts of native vegetation. Ridge-top sites are the most resilient in the face of suburban pollution, but are the first areas to be developed. Such remnants are often in sites planned for future sporting facilities, and become a source of conflict. As a result many of the rarer species in our Sydney area are those confined to ridge-top sites. For example, the rare *Prostanthera densa* just manages to survive at Bass and Flinders Point, Cronulla. Bushland in gullies too steep to build on is often retained, but these sites are most vulnerable to suburban run-off and pollution. Experience in older suburbs shows that these will become seriously degraded by weed invasion within the next 10 to 20 years. In sandstone areas, long-term viability of bushland can only be guaranteed by protection of whole sub-catchments from urban pollution. Residents, attracted to an area for its bushland character, often see their bushland deteriorating because inadequate provision has been made for its long-term survival.

The inexorable spread of the southern suburbs across the sandstone and shale landscapes stopped short of the sands of the Kurnell Peninsula. Here the landforms record some of the dramatic geomophological forces that have shaped Sydney's landscape. Three distinct land units are recognisable: an outlier of Hawkesbury Sandstone, partly covered by ancient podzolised wind-blown sand dunes, forms the rocky headlands and sea cliffs of Endeavour Heights; low-lying sand dunes and flats grade into mudflats, forming Towra Point, Bonna Point, and the shorelines of Weeney, Quibray and Woolooware Bays; and younger sand deposits form dunes behind the beaches of Bate Bay between Kurnell and Cronulla.

At the height of the last ice age, about 20,000 years ago, sea level was much lower and erosive

Early roads across the swamps and sand dunes of Kurnell had various obstacles. (Sutherland Shire Libraries and Information Service, Local Studies Collection)

forces carried much sand to the sea. As the climate warmed and the sea level rose, much of this sand was returned landwards, forming at one stage a sand barrier between Jibbon Head, on the southern side of Port Hacking, and the Kurnell headland. This was subsequently removed by marine currents north-west to form Bate Bay, trapping wind-blown sand and blocking the former mouth of the Georges River, which changed course to flow out through Botany Bay. Mud and sand were deposited as levees and deltaic fingers that coalesced to form Towra Point. Sea level stabilised about 6,000 years ago, but these landforms are still undergoing dynamic changes, due to natural processes and human impacts.

Over the last 200 years human activities have certainly reduced the area of native vegetation. Between 1815 and 1840 the Connell family cleared large trees from Kurnell and Woolooware. After Thomas Holt purchased the Kurnell Peninsula in the 1860s, major efforts were made to clear native vegetation using ring-barking and burning. Along the shores of Quibray Bay, 'the heavy timber of Honeysuckle [*Banksia*], Bang Alley [*Eucalyptus botryoides*], Oak [*Casuarina glauca*], and Gum [*Angophora costata*] is well killed by the Ring-barking,' reported his solicitor Walker, 'but having nothing done to it since then, the Scrub is now very thick, consisting principally of black wattle [*Acacia*], and a kind of Honeysuckle, and tea tree [*Leptospermum laevigatum*]'. Native shrubs proved very 'difficult and troublesome to get rid of'. *Themeda australis,* the Kangaroo Grass, and other native grasses were encouraged along with couch as feed for cattle, but the unpalatable Blady Grass, *Imperata cylindrica,* persisted on the sand hills. Walker describes a very large *Angophora costata* near the south-western shore of Woolooware Bay, nearly 7 m in circumference and 'capable of shading over 1000 sheep under its beautiful curly branches'[79].

Thomas Holt also had the mangroves and saltmarsh at the head of Gwawley Bay, near Taren Point, drained and reshaped into a series of canals for oyster culture. The venture failed, but in 1924 Frank Cridland speculated 'some day, perhaps, a water suburb will be laid out there with waterways instead of macadamized roads. Such an area of freehold tidal waters offers all sorts of novel town-planning possibilities and problems'[80]. His prophetic words have been realised in the canal estate development of Sylvania Waters.

After World War II, dunes were levelled and swamps filled to make way for a major oil refinery at Quibray Bay. Extraction of building sand destroyed heath, woodland and freshwater swamps behind Bate Bay. The sand hills near North Cronulla, once known as 'Green Hills Ridge', have been denuded and Connell Hill, once rising to 44 m, has been virtually demolished, trucked away for Sydney's buildings. The still-vegetated 'Calsil Dune', with its panoramic view, is not protected and could disappear. Dredging and reclamation for Sydney Airport and Port Botany on the northern side of Botany Bay have altered wave patterns, removed sand, and changed the Bay's shorelines.

Despite considerable change, significant bushland survives. We can still see much of the original diversity that compelled Banks and Solander to collect 'so many plants' during those eight days in 1770. On the rocky headlands of Endeavour Heights, heath, woodland and open-forest grow on the covering of ancient wind-blown sand dunes. Along the Yena and Muru tracks in Botany Bay National Park, Scribbly Gums *Eucalyptus haemastoma,* Red Bloodwood *Eucalyptus gummifera,* and smooth-barked *Angophora costata* are the dominant trees. The smaller plants include yellow-flowered *Hibbertia scandens, Dianella* lilies with bright blue fruits, fine-leaved *Astroloma pinifolium,* and Grass Trees, *Xanthorrhoea.* Shrubby *Leptospermum laevigatum, Banksia integrifolia, Acacia longifolia,* and *Persoonia lanceolata* are abundant. In the sheltered swales grow *Elaeocarpus reticulatus, Ceratopetalum gummiferum, Glochidion ferdinandi, Pittosporum revolutum,* and *Smilax glyciphylla* with its clusters of black fruit. The Woody Pear tree, *Xylomelum pyriforme,* and a shrubby yellow-brown pea, *Daviesia mimosoides,* are common beside the tracks.

On cliff-tops facing the sea, wind-pruned plants of *Westringia fruticosa, Leptospermum laevigatum, Baeckea imbricata* and *Acacia longifolia* tolerate the salt spray.

Correa alba and the Flannel Flower, *Actinotus helianthi,* cling to rock faces, while white-flowered *Pimelea linifolia* and pink-flowered *Eriostemon buxifolius* shelter behind larger plants. The track to Cape Baily crosses these heath-covered heights, offering superb views from the exposed cliff-tops. Here dense thickets of *Allocasuarina distyla, Banksia ericifolia* and *Hakea teretifolia* are interspersed with more open heath, coloured by the yellow flowers of *Isopogon anemonifolius* and *Melaleuca nodosa.* Around freshwater rock pools the distinctive red and white tubular flowers of *Epacris longiflora,* the vivid red bottlebrushes of *Callistemon citrinus,* and the minute-leaved *Baeckea brevifolia,* appear amongst clumps of *Baumea* and *Phragmites australis.* The older sand dunes south of Tabbigai Gap carry low coastal heath and scrub with similar plants to those of the rocky plateaus. In the swales between the dunes, in small freshwater swamps, grow water plants, yellow-flowered *Philydrum lanuginosum* and *Villarsia exaltata* with its upright spade-shaped leaves, and sedges *Leptocarpus tenax, Gahnia* and *Scirpus.* Similar freshwater swamps occur in the dunes behind Boat Harbour, and near the oil refinery.

The shallow waters of Quibray and Woolooware Bays contain beds of seagrass. These are not marine algae or seaweeds, but flowering plants with long strap-like leaves that grow below low tide level. *Zostera capricorni,* Eelgrass, grows closer inshore, while *Posidonia australis* grows in water up to 3 m deep. Seagrass beds contribute to the maintenance of the estuarine ecosystem by providing shelter and nourishment for fish and other marine life.

The intertidal zone is dominated by mangroves, the Grey Mangrove, *Avicennia marina,* with its twisted grey trunks and peg-like root pneumatophores, projecting above the muddy surface, and the smaller more shrubby River Mangrove, *Aegiceras corniculatum.* Saltmarshes, low growing herbaceous communities, occupy sites inundated less frequently. Samphire, *Sarcocornia quinqueflora,* a small herb with cylindrical succulent stems, often tinged with red, is the most abundant species, though 12 saltmarsh species have been recorded on Towra Point. *Casuarina glauca,* Swamp Oak, forms patches of forest just above the tidal limit. In some places these are associated with freshwater wetland species, including a variety of sedges and rushes; elsewhere terrestrial species form the understorey. Reed swamps dominated by *Phragmites australis,* rushlands with *Typha orientalis,* or the uncommon *Scirpus litoralis,* and tea-tree swamp with dense thickets of fine-leaved *Melaleuca ericifolia* occur. A number of weeds have invaded the Swamp Oak forests, especially near Captain Cook Drive; these include Lantana, Castor Oil Plant, Wandering Jew, Honeysuckle and Kikuyu.

Of special interest are the small patches of littoral rainforest on Towra Point. These are dominated by small *Cupaniopsis anacardioides* trees, with several dozen other native species, including *Cayratia clematidea* and *Cissus antarctica* vines and small trees of *Acmena smithii, Ficus rubiginosa* and *Syzygium oleosum.* Lantana sometimes intrudes at the edges.

Mangrove and saltmarsh, small areas of freshwater swamp, *Casuarina* fringe forest, woodlands of Bangalay, Coast Banksia and Coastal Tea-tree, and small areas of rare littoral rainforest, are now conserved within the Towra Point Nature Reserve, the most important estuarine nature reserve in Sydney. Its saltmarshes are the last of reasonable size surviving, and the areas of freshwater wetland, though small, are important, as such habitats are now extremely rare in the region. However, Towra Point's vegetation has been threatened repeatedly. Clearing for grazing, proposals for ports, airports, canal estates and polluting industrial development have all been tried. It has only been by good fortune and the hard work of people with foresight and commitment that we still have this rich assemblage of plant communities. It seems that such areas only become widely appreciated as their very survival becomes jeopardised by alienation, pollution and fragmentation. It is ironical that bushland now seen to be so valuable — for education, recreation, species conservation, and as a haven for international migratory waterbirds — has had to be fought for so bitterly against short-term development interests. Other bushland reserves have had similar histories of conflict. It is important that decisions on the future of bushland areas give adequate weight to its long-term value. It is easy to be blinded by short-term interests that will ultimately be paid for by loss of long-term environmental quality.

6 Sydney's Bushland: The Future

The same or different?

The story of Sydney's bushland is of plants and landscapes shaped by millions of years of geological and climatic events, thousands of years of Aboriginal occupation, and a mere two centuries of European impact. Today the area of bushland in the County of Cumberland has been reduced to less than half of its former coverage, though most of what remains is on the sandstone landscapes, areas of little use for productive agriculture, and initially too rugged and remote for suburban houses.

Sydney's other types of vegetation have not fared so well, being cleared for agricultural lands or suburbs at a time when bushland seemed a limitless resource. Today, less than 1% of the original Blue Gum High Forest remains, and less than 0.5% of the Turpentine-Ironbark Forest. Together these two communities once covered 46,000 ha. Barely 3% of each of the original 19,000 ha of River-flat Forest and 8,000 ha of Eastern Suburbs Banksia Scrub remain. Of the Cumberland Plain Woodlands, today's fragments add up to only 6% of the former 107,000 ha. About 5,000 ha of Castlereagh Woodland remains, but much of this is fragmented and very little is adequately conserved. About 1700 ha of estuarine wetlands remain, fortunately much of it conserved at Towra Point, but of the 540 ha of freshwater swamp, very little is protected[1]. Today's growing appreciation of bushland values and increased knowledge of the extent of our impact on the different vegetation types allow us to see bushland as a valuable but limited resource that cannot be replaced once destroyed.

What then is the future for our suburban bushland?

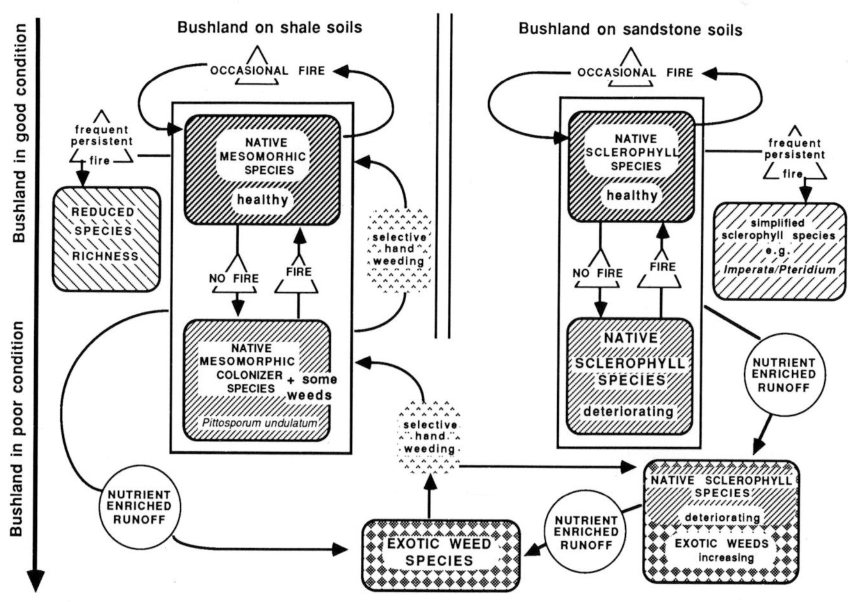

Diagram illustrating interactions between vegetation, nutrients, weeds and fire in bushland on Sydney's higher rainfall shale and sandstone soils.

Continuing depletion?

Will the areas of bushland continue to shrink in the near future? Despite environmental protection policies designed to conserve urban bushland, a fair proportion is still unprotected. Some of the pressures which will continue to threaten bushland are the sale of public land for housing, the widening of roads, construction of freeways, and the channelling or piping of creeks. For example, bushland on Commonwealth land at Long Bay and Mosman has been proposed for housing, while freeway proposals threaten bushland at Wolli Creek and in the Lane Cove River valley. Near Riverstone an important proposed nature reserve is under threat of disturbance by power and water service corridors.

Bushland in Sydney is now too scarce to be regarded as vacant land awaiting a use. On the Cumberland Plain in western Sydney, the bush on the flat shale country has been undervalued in comparison with the scenic bushland of the nearby Blue Mountains. Virtually none is protected in conservation reserves. The remnants are vulnerable, and as the surrounding freehold land has often been cleared for farming, are limited in extent, often confined to publicly owned Crown land. Pressures to use these bushland remnants as residential land to solve short-term housing crises or as revenue-raising exercises must be resisted.

Bushland remnants provide the future urban bushland for new suburbs, and should be integrated with the planned development of adjacent cleared areas. In the past, areas were often left as bush, with the long-term aim of clearing them for sporting ovals, golf courses, future freeways, etc. These bushland areas are often assumed by residents to be permanently protected, and there is alarm and ill-feeling when they are destroyed. 'Local bush' for new suburbs needs to be accepted as a legitimate land use and provided for in subdivision planning, just as ovals should be. Failure to do this will lead to future conflict as residents seek to preserve the amenity which attracted them to their suburb. The value of bushland in the older suburbs is now appreciated. We must ensure that similar bushland is a planned part of the newly developing suburbs.

An irreplaceable resource?

Sydney's bushland is scarce; there is no more being made, indeed it cannot be made — it is not just a garden, where plants depend on human intervention for their propagation and survival. Rather, it is an assemblage of interacting plant populations, each species responding to a variety of influences and able to continue its life cycle without human interference. To recreate this diversity requires the establishment of perhaps 100 plant species in a small area. In landscaping projects, where attempts have been made to re-make bushland, the best usually achieved is a replanting of trees and shrubs, restoring only a fraction of the former diversity. Many shrub, vine, grass and herb species are overlooked. These are an important component of natural communities, but propagation or establishment of self-perpetuating populations may be impossible. In addition, once the land has been used for other purposes and soil and drainage patterns have been disturbed, many plants are unable to establish new generations naturally. Even where topsoil has been saved for major revegetation projects, many species have failed to reappear. The natural cycle is hard to start. Inevitable competition from weeds, the absence of fire, and changes to the soils militate against success. Eucalypts and mown grass do not make a natural plant community.

Deteriorating bushland, however, can be rehabilitated, depending on the changes that have been incurred in the system, particularly relating to the soil conditions. Where nutrient is being added in urban run-off, mesic species can be encouraged, but weeds will be likely to predominate without continuing maintenance. Bush regeneration programs that concentrate on the selective removal of exotic weeds and the encouragement of native species have developed from the work of Joan and Eileen Bradley and the National Trust of Australia (NSW), and provide an important way of dealing with localised weed problems in bushland parks.

Changes in species composition

Will bushland reserves as we know them remain in 50 or 100 years, or will they be weedy thickets?

Some will and some won't. Our evidence shows that the successful exotic weed species invade bushland following disturbance generally associated with increases in the nutrient content of the systems. Gullies receiving urban run-off are particularly susceptible, and it is highly likely that most gullies, except those with protected or internal catchments,

Fig. 19 Use of Sydney flower designs for decoration in the early twentieth century. (Private collection)

Fig. 20 In 1865 von Guerard depicted extensive areas of heath, scrub and woodland extending from Old South Head Road to South Head. (*Sydney Heads 1865*, Eugene von Guerard, Art Gallery of New South Wales)

Fig. 21 Fire is being used experimentally to promote natural regeneration in this remnant of Blue Gum High Forest in Sheldon Forest at Pymble. (1989)

Exotic species invade bushland where there has been disturbance, and are often particularly conspicuous along bushland margins. Here is *Lantana camara* on the edge of bushland at Ryde.

will become weed-dominated in due course. On the other hand, ridges, dry hillslopes and the flatter dry sites such as on the Cumberland Plain are not so vulnerable to those exotic species, and will never become weedy, or weeds will establish only at a slow rate so that infrequent weeding may be adequate to control them. We should remember, however, that many such sites are now quite weedy because of the absence of any control in the past. Even at a slow rate of establishment, the weeds can build up if they are given 50–100 years.

The effect of fire?

This is a difficult one. Recent studies into the effect of fire on native species are beginning to show its complexity. The fire frequency, temperature, season, prior and subsequent weather conditions, and proximity of nearby unburnt bush are all important. At the same time our options on the use of fire as a management tool in urban bushland are decreasing. Most areas are now too small or too near houses to ever support a hot crown fire, even an unplanned one, while the season and conditions for cool weather burns are becoming more restrictive. Indeed 'no burn' council areas are severely limiting the use of fire as a management tool.

In the absence of fire, bushland will survive, but particularly in moist sites it will become increasingly dominated by a smaller group of native species, notably *Pittosporum undulatum,* and indeed woody weeds such as Privet and Lantana. The eucalypts themselves will be affected as they require open conditions for establishment, and often open conditions for healthy growth.

In larger bushland areas there are pressures for frequent burning to reduce fuel, particularly on perimeters. Whilst many species regenerate vigorously from rootstocks or trunks, other species are killed. These species need a minimum fire-free period between burns to build-up their seed reserves. A fire frequency of less than this period will eliminate the species from the plant community. *Banksia ericifolia,* for example, needs at least eight years to mature and begin to produce seeds[2]. In frequently-burnt areas floristic composition will become simplified as seeders disappear and resprouting species such as Bracken Fern, *Pteridium esculentum,* and Blady Grass, *Imperata cylindrica,* are advantaged.

Interactions between vegetation, nutrients, weeds and fire can be expressed diagrammatically as shown here for bushland on Sydney's higher rainfall shale and sandstone soils. Models like this can be developed to incorporate the appropriate conditions for individual areas and provide a basis for understanding the implications of particular management strategies.

Have many species become extinct?

We will never be sure how many species, restricted in range and growing on the fertile lands, may have become extinct before they were recorded. In early colonial days the few botanists available to collect and describe local plants would have had a difficult task keeping ahead of clearing in the rapidly expanding settlement.

Today the number of species remaining in the Sydney area is about 1,500, although populations of many have been severely reduced. In the list of nationally rare and endangered plants prepared in 1988, one Sydney species was declared extinct. This was *Hypsela sessiliflora,* recorded from South Creek last century and now almost certainly gone. An undescribed species of *Persoonia* from the Warringah peninsula is now also considered extinct. Other species now extinct in the Sydney area, though still occurring elsewhere in the country, include *Swainsona monticola, Bauera capitata, Tylophora woollsii, Tetratheca juncea, Hibbertia virgata* and *Acacia quadrilateralis.* Despite the enormous loss of area of bushland, the number of extinctions is so far relatively small, although many local populations have disappeared and many more, particularly in western Sydney, are poised on the brink. Species at greatest risk belong to plant communities that have virtually disappeared.

Of the Sydney flora, about two-thirds are sandstone species and likely to be conserved. However, it is the species of the shale and river flats that are most vulnerable. Many of these are known from a number of places now, but with the rate of change in western Sydney will disappear in the next 20 years. Changes such as the use of herbicides will increase the rate of loss. Because many are also found outside the Sydney area, they are not regarded as threatened, though since similar processes are occurring, many will disappear there at the same time.

What should we do?

Firstly, we should take urgent steps to conserve the significant areas of bushland still remaining in Sydney's west. At present there are virtually no major nature reserves though some western Sydney councils are helping on a small scale. The most significant of the remaining areas are:

- The catchment of Longneck Lagoon which includes important areas of Grey Box woodland and protects the catchment of this important wetland (see Hawkesbury);
- The southern additions to Agnes Banks Nature Reserve; these would protect significant areas of Castlereagh Woodland on poorly drained sites (see Penrith);
- Castlereagh State Forest, which needs permanent protection as a flora reserve for its important ironbark forests and diverse flora (see Penrith);
- Commonwealth land at Shanes Park, which should be acquired to protect intergrading vegetation on shale and Tertiary Alluvium (see Blacktown);
- The Mulgoa Nature Reserve proposal with its interesting shale cliff-line habitat (see Penrith);
- The Kemps Creek Nature Reserve proposal, which should be established to protect low-lying Cumberland Plain Woodland (see Liverpool);
- Water Board land around Prospect Reservoir, which needs permanent protection for its important woodlands (see Blacktown).

These areas include the last remaining stands of native bushland in western Sydney and will serve as key points in a network of reserves as the spaces in between fill with houses. Other smaller areas will hopefully be looked after by councils and these, together with tree-planting schemes, will make corridors linking these major reserves. **But we must create the large reserves before it is too late.** Most of these areas are Crown land within public ownership and could be readily protected.

Significant areas closer to Sydney that should be protected have been mentioned throughout this book, but special attention should be given to bushland at Silverwater (see Auburn), Twin Road (see Ryde), Jennifer Street and Long Bay (see Randwick), Wolli Creek (see Canterbury) and Lansdowne (see Bankstown).

Secondly, research has shown that bushland deterioration is closely related to changes in catchments and stormwater run-off. We must use this knowledge when we design new suburban subdivisions. Bushland should be identified as significant and its catchments and subcatchments protected accordingly. We should not build on every ridge and so pollute every stream. Drainage from housing should have well-maintained silt traps as a minimum and should not be directed straight into bushland.

Thirdly, we should ensure that bushland already conserved does not deteriorate. A particular threat in smaller parks is the over-zealous use of the

Plant species diversity reduced tenfold, and habitat for native birds removed, by over-zealous use of lawnmowers in part of Mirambeena Regional Park at Lansdowne. (1989)

In contrast, the healthy understorey of shrubs, herbs and seeding native grasses provides a diversity of habitats for native animals and birds in nearby Lansdowne Park. (J. Plaza, RBG, 1990)

lawnmower and whipper-snipper around bushland margins to give a 'tidy' edge. The perception that bushland should look tidy denies its natural and ever-changing nature. Over-tidying of nature shows a lack of understanding; its effect on bushland may be insidious but can be devastating over time. Even in 'tidy' Britain this destructiveness has been deplored:

More intractable than destruction in pursuit of a purpose. . . is the blight of tidiness which every year sweeps away something of beauty or meaning. . . destroying ivy-tods and 'misshapen' trees, cutting hedges to the ground every year, devouring saplings, levelling churchyards, filling ponds, pottering with paraquat — all the little, often unconscious vandalisms that hate what is tangled and unpredictable but create nothing.[3]

Around Sydney evidence of similar attitudes is rife — lawn-mowers, whipper-snippers and herbicide sprays are applied conscientiously to tidy up stragglers at bushland edges. Bit by bit the bush is pushed back and the Kikuyu invades. Examples abound of bushland destroyed by a sequence of small, seemingly reasonable individual actions — road widening, drain construction, tree lopping, firebreak clearing, edge slashing, tidying up of dumped rubbish — the 'tyranny of small decisions'[4]. This destruction is so unnecessary — for example, physical barriers such as logs can be used to protect bushland from the ever-increasing incursions of lawn-mowers. Years ago Walter Burley Griffin recognised the maintenance-free nature of our bushland. He described it as

the cleanest most delicate and varied native ligneous evergreen perpetually blooming flora extant. . . For these reasons no vegetation could be better to live with, free as it is of rank growths, brambles, nettles, burrs, weeds or plants seasonably untidy. Moreover it will persist through drought without watering and recover from abuse without help provided the peculiar nature of the soil is respected and manure with exotic seed is kept out.[5]

To protect and maintain our bushland requires the actions of governments and the vigilance of individuals. The role of bushland in enriching our lives and enhancing our suburbs, as well as for its own intrinsic value, is beginning to be generally appreciated. History shows that if our bushland is taken for granted, it will be quietly taken from us.

Appendix

Common Species of Sydney's Vegetation Types

Listed below for each vegetation type are up to 50 of the most commonly found species. Not all species will necessarily be found at any one site.

BGHF	Blue Gum High Forest
TIF	Turpentine-Ironbark Forest
CPW	Cumberland Plain Woodlands
CW	Castlereagh Woodlands
RFF	River-flat Forest
SHW	Sandstone Heaths, Woodlands and Forests
ESB	Eastern Suburbs Banksia Scrub
FW	Freshwater Wetlands
E	Estuarine Wetlands

	BGHF	TIF	CPW	CW	RFF	SHW	ESB	FW/E
Ferns								
Adiantum aethiopicum	BGHF				RFF			
Blechnum cartilagineum	BGHF							
Cheilanthes sieberi		TIF	CPW	CW				
Culcita dubia	BGHF							
Doodia aspera					RFF			
Hypolepis muelleri					RFF			
Lindsaea linearis						SHW		
Pteridium esculentum	BGHF				RFF	SHW	ESB	
Monocotyledons								
Alisma plantago-aquatica								FW
Anisopogon avenaceus						SHW		
Aristida ramosa			CPW					
Aristida vagans		TIF	CPW	CW				
Arthropodium milleflorum			CPW					
Baumea juncea								E
Caustis pentandra							ESB	
Chloris ventricosa			CPW					
Commelina cyanea			CPW					
Cyathochaeta diandra				CW		SHW		
Cyperus eragrostis								E
Cyperus gracilis			CPW					
Dianella caerulea		TIF				SHW		
Dianella laevis			CPW					
Dianella revoluta			CPW	CW			ESB	
Dichelachne micrantha			CPW					
Dichelachne rara		TIF						
Diuris maculata				CW				
Echinopogon caespitosus	BGHF	TIF	CPW					
Echinopogon ovatus			CPW					
Elaeocharis sphacelata								FW
Entolasia marginata			CPW					
Entolasia stricta		TIF		CW		SHW		
Eragrostis brownii				CW			ESB	
Eragrostis leptostachya			CPW					
Eustrephus latifolius	BGHF				RFF			
Haemodorum planifolium								ESB
Hypolaena fastigiata							ESB	
Hypoxis hygrometrica			CPW					
Imperata cylindrica	BGHF	TIF						
Isolepis nodosus								E
Juncus kraussii								E
Juncus prismatocarpus								FW
Juncus subsecundus				CW				
Juncus usitatus								FW
Lepidosperma laterale		TIF	CPW	CW			ESB	
Leptocarpus tenax							ESB	
Lepyrodia scariosa						SHW	ESB	
Lomandra cylindrica						SHW		
Lomandra filiformis			CPW	CW				
Lomandra longifolia	BGHF	TIF			RFF	SHW		
Lomandra multiflora			CPW					
Lomandra obliqua						SHW		
Microlaena stipoides		TIF	CPW					
Oplismenus aemulus			CPW					
Panicum simile		TIF	CPW					
Patersonia glabrata						SHW		
Patersonia sericea				CW		SHW		
Poa affinus	BGHF							
Ptilanthelium deustum				CW				
Schoenus apogon								E
Schoenoplectus validus								E
Scirpus prolifer								FW
Smilax glyciphylla	BGHF					SHW		
Sporobolus virginicus								E
Stipa pubescens		TIF						
Themeda australis	BGHF	TIF	CPW	CW				
Tricoryne elatior			CPW					
Triglochin procera								FW
Triglochin striata								E
Xanthorrhoea media						SHW		
Xanthorrhoea minor				CW				
Xanthorrhoea resinosa							ESB	
Zoysia macrantha								E
Dicotyledons								
Acacia brownii				CW				
Acacia decurrens			CPW					
Acacia elongata				CW				
Acacia falcata		TIF		CW				
Acacia filicifolia					RFF			
Acacia floribunda					RFF			
Acacia implexa	BGHF		CPW					
Acacia longifolia		TIF				SHW	ESB	
Acacia myrtifolia		TIF						
Acacia parramattensis		TIF	CPW		RFF			
Acacia suaveolens		TIF				SHW	ESB	
Actinotus helianthi						SHW		

	BGHF	TIF	CPW	CW	RFF	SHW	ESB	FW/E
Actinotus minor						SHW	ESB	
Aegiceras corniculatum								E
Allocasuarina distyla							ESB	
Allocasuarina littoralis								
Allocasuarina torulosa	BGHF	TIF						
Alphitonia excelsa					RFF			
Angophora bakeri				CW				
Angophora costata	BGHF					SHW		
Angophora floribunda	BGHF				RFF			
Angophora hispida						SHW		
Angophora subvelutina					RFF			
Apium prostratum								E
Asperula conferta				CW				
Astroloma pinifolium							ESB	
Avicennia marina								E
Backhousia myrtifolia					RFF			
Banksia aemula							ESB	
Banksia ericifolia						SHW	ESB	
Banksia integrifolia							ESB	
Banksia serrata						SHW		
Banksia spinulosa				CW				
Billardiera scandens	BGHF	TIF				SHW		
Baeckea imbricata							ESB	
Bauera rubioides							ESB	
Boronia parviflora							ESB	
Boronia rigens							ESB	
Bossiaea heterophylla							ESB	
Brachycome angustifolia	BGHF							
Brachyloma daphnoides							ESB	
Breynia oblongifolia	BGHF	TIF			RFF			
Brunoniella australis			CPW	CW				
Bursaria spinosa		TIF	CPW	CW	RFF			
Cassytha glabella				CW				
Casuarina cunninghamiana					RFF			
Casuarina glauca								E
Ceratopetalum gummiferum						SHW		
Centella asiatica		TIF						
Clematis aristata	BGHF							
Clematis glycinoides	BGHF	TIF						
Clerodendrum tomentosum	BGHF							
Cotula coronopifolia								E
Dampiera stricta						SHW		
Darwinia fascicularis							ESB	
Darwinia leptantha							ESB	
Daviesia ulicifolia		TIF		CW				
Desmodium rhytidophyllum	BGHF							
Dichondra repens		TIF	CPW					
Dillwynia juniperina			CPW					
Dillwynia retorta						SHW		
Dillwynia tenuifolia				CW				
Dodonaea triquetra		TIF				SHW		
Duboisia myoporoides					RFF			
Elaeocarpus reticulatus	BGHF					SHW		
Epacris microphylla							ESB	
Epacris obtusifolia							ESB	
Eucalyptus amplifolia					RFF			
Eucalyptus crebra			CPW					
Eucalyptus deanei					RFF			
Eucalyptus eugenioides			CPW					
Eucalyptus fibrosa			CPW	CW				
Eucalyptus globoidea		TIF						
Eucalyptus gummifera		TIF				SHW	ESB	
Eucalyptus haemastoma		TIF				SHW		
Eucalyptus moluccana			CPW					
Eucalyptus paniculata	BGHF	TIF						
Eucalyptus parramattensis				CW				
Eucalyptus pilularis	BGHF							
Eucalyptus piperita						SHW		
Eucalyptus resinifera		TIF						
Eucalyptus saligna	BGHF							
Eucalyptus sclerophylla				CW				
Eucalyptus tereticornis			CPW		RFF			
Exocarpos cupressiformis		TIF	CPW					
Glochidion ferdinandi					RFF			
Glycine clandestina			CPW					
Glycine tabacina	BGHF	TIF						
Gompholobium minus				CW				
Gonocarpus tetragynous				CW				
Gonocarpus teucrioides						SHW		
Goodenia belledifolia				CW				
Goodenia hederacea		TIF	CPW	CW				
Goodenia heterophylla	BGHF							
Grevillea buxifolia						SHW		
Grevillea linearifolia						SHW		
Grevillea mucronulata				CW				
Hakea dactyloides				CW				
Hakea sericea				CW				
Hakea teretifolia							ESB	
Hardenbergia violacea	BGHF	TIF	CPW	CW			ESB	
Helichrysum diosmifolium		TIF		CW				
Helichrysum scorpioides	BGHF							
Hibbertia diffusa			CPW					
Hibbertia fasciculata							ESB	
Hibbertia linearis						SHW		
Hibbertia obtusifolia							ESB	
Hibbertia scandens	BGHF							
Hypericum gramineum			CPW	CW				
Indigofera australis			CPW					
Kennedia rubicunda	BGHF	TIF						
Kunzea ambigua		TIF				SHW	ESB	
Lambertia formosa						SHW	ESB	
Leptospermum attenuatum				CW		SHW	ESB	
Leptospermum flavescens					RFF			
Leptospermum laevigatum							ESB	
Leucopogon juniperinus	BGHF							
Leucopogon lanceolatus	BGHF							
Lissanthe strigosa				CW				
Lobelia alata								E
Lomatia silaifolia						SHW		
Melaleuca decora		TIF	CPW	CW				
Melaleuca linariifolia								FW
Melaleuca nodosa			CW					
Melaleuca squamea							ESB	
Micrantheum ericoides						SHW		

	BGHF	TIF	CPW	CW	RFF	SHW	ESB	FW/E
Mitrasacme polymorpha				CW		SHW		
Monotoca elliptica							ESB	
Monotoca scoparia							ESB	
Myoporum debile			CPW					E
Myoporum insulare								E
Notelaea longifolia	BGHF	TIF						
Omalanthus populifolius					RFF			
Opercularia diphylla				CW				
Oxalis exilis	BGHF	TIF	CPW					
Pandorea pandorana	BGHF	TIF						
Persicaria hydropiper								FW
Persicaria strigosum								FW
Persoonia linearis	BGHF							
Persoonia lanceolata							ESB	
Philotheca salsolifolia							ESB	
Philydrum lanuginosum								FW
Phyllanthus thymoides						SHW		
Phyllanthus filicaulis			CPW					
Phyllanthus gasstroemii					RFF			
Phyllota phylicoides						SHW		
Pimelea linifolia				CW		SHW	ESB	
Pittosporum revolutum	BGHF							
Pittosporum undulatum	BGHF	TIF				SHW		
Platylobium formosum	BGHF							
Platysace ericoides				CW				
Platysace linearifolia						SHW		
Polyscias sambucifolia	BGHF	TIF				SHW		
Pomax umbellata		TIF		CW				
Poranthera microphylla	BGHF							
Pratia purpurascens	BGHF	TIF	CPW	CW				
Pseuderanthemum variabile	BGHF	TIF						
Pultenaea elliptica				CW		SHW		
Ranunculus inundatus								FW
Rapanea variabilis	BGHF							
Rubus parvifolius	BGHF							
Samolus repens								E
Sarcocornia quinqueflora								E
Solanum pungetium			CPW					
Stephania japonica					RFF			
Stylidium graminifolium				CW				
Styphelia viridis							ESB	
Syncarpia glomulifera	BGHF	TIF						
Suaeda australis								E
Tetragonia tetragonioides								E
Trema aspera					RFF			
Tristaniopsis laurina					RFF			
Tylophora barbata	BGHF							
Typha orientalis								FW
Vernonia cinerea			CPW					
Veronica plebeia		TIF						
Wahlenbergia gracilis			CPW					
Wilsonia backhousei								E
Woollsia pungens						SHW		
Xanthosia pilosa						SHW		
Zieria smithii	BGHF							

References

References not specifically cited but providing further detail are marked with asterisks.

Chapter 1: Sydney's Landscape

Geology

*Herbert, C. (ed.) *Geology of the Sydney 1:100 000 Sheet 9130.* New South Wales Dept of Mineral Resources, Sydney, 1983.

*Herbert, C. & Helby, R. (eds) *A Guide to the Sydney Basin.* Geological Survey of N.S.W. Bulletin No. 26, New South Wales Dept of Mineral Resources, Sydney, 1980.

**Sydney 1:250 000 Geological Series Sheet.* 3rd edn, New South Wales Dept of Mines, Sydney, 1966.

Soils

*Chapman, G.A. & Murphy, C.L. *Soil Landscapes of the Sydney 1:100 000 Sheet.* Soil Conservation Service of N.S.W., Sydney, 1989.

Climate

1 Bureau of Meteorology, *Climatic Survey Sydney Region 5 New South Wales.* Australian Govenment Publishing Service, Canberra, 1979.

*Gentilli, J. *Australian Climate Patterns.* Thomas Nelson, Adelaide, 1972.

Chapter 2: Plants in Aboriginal Life

1 Kohen, J.L. & Lampert, R. Hunters and fishers in the Sydney region. In: *Australians to 1788* (D.J. Mulvaney & P.J. White, eds). Fairfax, Syme & Weldon Associates, Sydney, 1987.

2 Caley, George, quoted in *Reflections on the Colony of New South Wales: George Caley* (J.E.B. Currey, ed.). Lansdowne Press, Melbourne, 1966.

3 Bradley, William *A voyage to New South Wales: The Journal of Lieutenant William Bradley RN of HMS Sirius 1786-1792.* Facsimile edn. Trustees of the Public Library of N.S.W. & Ure Smith, Sydney, 1969.

4 Kohen, J.L. Computer analysis of Aboriginal-plant interactions on the Cumberland Plain. In: *Technology in the 80's. Proceedings of a Conference on Science Technology.* ANZAAS-AIST, Macquarie University, 1984.

5 Hunter, John *An historical journal of events at Sydney and at Sea 1787-1792* (first published London, 1793). Angus & Robertson, Sydney, 1968.

6 Maiden, J.H. *The Useful Native Plants of Australia* (first published Sydney, 1889). Facsimile edn. Compendium, Melbourne, 1975.

7 White, John *Journal of a voyage to New South Wales* (first published 1790). Angus & Robertson, Sydney, 1962.

8 Beaton, J.M. Fire and water: aspects of Australian Aboriginal management of Cycads. *Archaeology in Oceania* 17, 51-58, 1982.

9 Gill, A.M. & Ingwersen, F. Growth of *Xanthorrhoea australis* R. Br. in relation to fire. *Journal of Applied Ecology* 13, 195-203, 1976.

*Haigh, C. & Goldstein, W. The Aborigines of New South Wales. *Parks and Wildlife* 2(5), N.S.W. National Parks and Wildlife Service, Sydney.

*Turbet, P. *The Aborigines of the Sydney District before 1788.* Kangaroo Press, Sydney, 1989.

Chapter 3: Sydney's Vegetation Types

1 Specht, R.L. Vegetation. In: *The Australian Environment* (G.W. Leeper, ed.). 4th edn. CSIRO-Melbourne University Press, Melbourne, 1970.

2 Beadle, N.C.W., Evans, O.D. & Carolin, R.C. *Flora of the Sydney Region.* 3rd edn. A.H. & A.W. Reed, Sydney, 1982.

3 Fairley, A. & Moore, P. *Native Plants of the Sydney District: An Identification Guide.* Kangaroo Press, Sydney, 1989.

4 Adam, P. Saltmarsh plants of New South Wales. *Wetlands (Australia),* 1(1), 11-19, 1981.

5 Baker, M., Corringham, R. & Dark, J. *Native Plants of the Sydney Region.* Three Sisters Productions, Winmalee, 1986.

6 Child, J. *Trees of the Sydney Region.* Cheshire-Lansdowne, Melbourne, 1968.

7 Clarke, I. & Lee, H. *Name That Flower: The Identification of Flowering Plants.* Melbourne University Press, Melbourne, 1987.

8 Edmonds, T. & Webb, J. *Sydney Sandstone Flora.* New South Wales University Press, Kensington, 1986.

9 Harden, G. & Williams, J. *How to Identify Plants.* University of New England, Armidale, 1979.

10 Harden, G.J. (ed.) *Flora of New South Wales.* Volume 1. New South Wales University Press, Kensington, 1990.

11 Robinson, L. *Native Trees of Sydney.* Gould League of NSW, Sydney, 1988.

12 Rotherham, E.R., Briggs, B.G., Blaxell, D.F. & Carolin, R.C. *Flowers and Plants of New South Wales and Southern Queensland.* A.H. & A.W. Reed, Sydney, 1975.

13 Cunningham, P. *Two years in New South Wales.* London, 1827.

14 Meredith, Mrs Charles *Notes and sketches of New South Wales* (first published London, 1844). Facsimile edn. Penguin Books, Harmondsworth, 1973.

15 Phillip, Arthur *The voyage of Governor Phillip to Botany Bay* (first published London, 1789). Facsimile edn. Hutchinson Group (Australia), Melbourne, 1982.

16 Atkinson, James *An account of the state of agriculture and grazing in New South Wales* (first published London, 1826). Facsimile edn Sydney University Press, Sydney, 1975.

17 Cunningham, Allan, quoted in Ida Lee, *Early explorers in Australia.* Methuen, London, 1925.

18 Macarthur to King 12/10/1805, *Historical Records of Australia* 5, 577.

19 Mossman, Samuel & Banister, Thomas *Australia visited and Revisited,* (first published London, 1853). Facsimile edn. Ure Smith, Sydney, 1974.

20 Hunter, John *An historical journal of events at Sydney and at sea, 1787-1792* (first published London, 1793). Angus & Robertson, Sydney, 1968.

21 Woolls, W.W. Eucalypts of the County of Cumberland Part III. *Proceedings of the Linnean Society of New South Wales* 5, 463-469, 1880.

22 Hamilton, A.A. An ecological study of the saltmarsh vegetation in the Port Jackson district. *Proceedings of the Linnean Society of New South Wales* 44, 463-513, 1919.

23 Tench, Watkin *Sydney's First Four Years: Being a reprint of A Narrative of the Expedition to Botany Bay and A Complete Account of the Settlement at Port Jackson.* Library of Australian History, Sydney, 1979.

24 Bradley, William *A voyage to New South Wales: The Journal of Lieutenant William Bradley RN of HMS Sirius 1786-1792.* Facsimile edn. Trustees of the Public Library of N.S.W. & Ure Smith, Sydney, 1969.

25 Hamilton, A.A. Topographical, ecological, and taxonomic notes on the ocean shoreline vegetation of the Port Jackson district. *Journal and Proceedings of the Royal Society of New South Wales* 51, 287-355, 1918.

26 McLoughlin, L. *The Middle Lane Cove River, A History and a Future.* Monograph No. 1, Centre for Environmental and Urban Studies, Macquarie University, 1985.

*Kartzoff, M. *Nature and a City : The Native Vegetation of the Sydney Area.* Edwards & Shaw, Sydney, 1969.

*Pickard, J. Annotated bibliography of floristic lists of New South Wales. *Contributions from the New South Wales National Herbarium* 4, 291-317, 1972.

*Bryant, H.J. & Benson, D.H. Recent floristic lists of New South Wales. *Cunninghamia* 1(1), 59-77, 1981.

*Keith, D.A. Floristic lists of New South Wales (III). *Cunninghamia* 2(1), 39-73, 1988.

*Pidgeon, I.M. The ecology of the Central Coastal area of New South Wales. I. The environment and general features of the vegetation. *Proceedings of the Linnean Society of New South Wales* 62, 315-340, 1937.

*Pidgeon, I.M. The ecology of the Central Coastal area of New South Wales. II. Plant succession on the Hawkesbury Sandstone. *Proceedings of the Linnean Society of New South Wales* 63, 1-26, 1938.

*Pidgeon, I.M. The ecology of the Central Coastal area of New South Wales. III. Types of primary succession. *Proceedings of the Linnean Society of New South Wales* 65, 221-249, 1940.

*Pidgeon, I.M. The ecology of the Central Coastal area of New South Wales. IV. Forest types on soils from Hawkesbury

Sandstone and Wianamatta Shale. *Proceedings of the Linnean Society of New South Wales* 66, 113-137, 1941.

Chapter 4: The European Impact

Early botanical work in Sydney

1 Beaglehole, J.C. (ed.) *The* Endeavour *journal of Joseph Banks 1768-1771.* Volume 2. Angus & Robertson, Sydney, 1962.
2 Phillip, Arthur, quoted in Gilbert, L. *The Royal Botanic Gardens, Sydney: A history 1816-1985.* Oxford University Press, Melbourne, 1986.
3 White, John *Journal of a voyage to New South Wales* (first published 1790). Angus & Robertson, Sydney, 1962.
4 Considen, Denis quoted in Maiden, J.H. *Sir Joseph Banks: the 'Father of Australia'.* Government Printer, Sydney, 1909.
5 Hooker, J.D. quoted in Hall, N. *Botanists of the Eucalypts* CSIRO, Melbourne, 1978.
6 Field, Barron Botany-Bay Flowers. In: *Geographical memoirs of New South Wales* (Barron Field, ed.). John Murray, London, 1825.

The impact of agriculture

7 Banks, Joseph, quoted in Gilbert, L. *The Royal Botanic Gardens, Sydney: a history 1816-1985.* Oxford University Press, Melbourne, 1986.
8 Worgan, G.B. *Journal of a First Fleet Surgeon.* Publication Number 16, The William Dixson Foundation. The Library Council of New South Wales in association with Library of Australian History, Sydney, 1978.
9 Tench, Watkin *Sydney's First Four Years: Being a reprint of A Narrative of the Expedition to Botany Bay and A Complete Account of the Settlement at Port Jackson.* Library of Australian History, Sydney, 1979.
10 Phillip to Sydney 15/5/1788, *Historical Records of Australia* 1, 19.
11 Ross 16/11/1788 quoted in King, J. *The First Settlement.* Macmillan, Melbourne, 1984.
12 Phillip to Sydney 28/9/1788, *Historical Records of Australia* 1, 73.
13 Field, Barron Journal of an excursion across the Blue Mountains of New South Wales. In: *Geographical Memoirs of New South Wales* (Barron Field, ed.). John Murray, London, 1825.
14 King to Hobart 4/10/1803, *Historical Records of Australia* 5, 67.
15 Atkinson, James *An account of the State of Agriculture and grazing in New South Wales.* (first published London, 1826. Facsimile edn University Press, Sydney, 1975.)
16 Meredith, Mrs Charles *Notes and sketches of New South Wales* (first published London, 1844). Facsimile edn. Penguin Books, Ringwood, 1973.

Changing attitudes

17 Darwin, Charles A journey to Bathurst in January, 1836. In: *Fourteen journeys over the Blue Mountains of New South Wales 1813-1814* (G. Mackaness, ed.). Part III. Review Publications, Dubbo, 1978.
18 Gibbs, Shallard & Co. *An Illustrated Guide to Sydney 1882.* Facsimile edn. Angus & Robertson, Sydney, 1981.
19 Gibbs, May *Flannel Flowers and other Bush Babies.* Angus & Robertson, Sydney, 1983.

'the overflow of bricks and mortar'

20 Woolls, William quoted in Gilbert, L. *William Woolls, 1814-1893 'a most useful colonist'.* Mulini Press, Canberra, 1985.
21 *The Railway Guide of New South Wales.* Government Printer, Sydney, 1879.
22 Quoted in Birch, A. & Macmillan, D.S. *The Sydney Scene 1788-1960.* Hale & Ironmonger, Sydney, 1982.

Cyclic and irreversible change

23 Clements, A. Suburban development and the resultant changes in the vegetation of the bushland of the northern Sydney region. *Australian Journal of Ecology* 8, 307-319, 1983.
24 Wright, H. The longterm threat to bushland from urban runoff — minimising the damage. In: *Caring for Warringah's Bushland.* Warringah Shire Council, 1988.
25 Leishman, M.R. Suburban development and resultant changes in the phosphorous status of soils in the area of Ku-ring-gai, Sydney. *Proceedings of the Linnean Society of New South Wales* 112, in press 1990.
26 Wright, Harley, personal communication, 1990.
27 Bliss, P.J., Riley, S.J. & Adamson, D. Towards rational guidelines for urban stormwater disposal into flora preservation areas. *The Shire and Municipal Record* 76, 181-185, 191, 1983.
28 Bradstock, R. & Myerscough, P.J. Fire effects on seed release and the emergence and establishment of seedlings of *Banksia ericifolia* L.f. *Australian Journal of Botany* 29, 521-531, 1981.
29 Benson, D.H. Maturation periods for fire-sensitive shrub species in Hawkesbury Sandstone vegetation. *Cunninghamia* 1(3), 339-349, 1985.
30 Siddiqi, M.Y., Carolin, R.G. & Myerscough, P.J. Studies in the ecology of coastal heath in New South Wales. III. Regrowth of vegetation after fire. *Proceedings of the Linnean Society of New South Wales* 101, 53-63, 1976.
31 Nieuwenhuis, A. The effect of fire frequency on the sclerophyll vegetation of the West Head, New South Wales. *Australian Journal of Ecology* 12, 373-385, 1987.
32 Pratten, C.H., personal communication, January 1990.
*Bradstock, R. & Fitzhardinge, R. *Bush in the City: A Report on Urban Bushland in the Sydney Metropolitan Area. 2. Hornsby-Upper North Shore & Lane Cove Valley.* Nature Conservation Council of N.S.W., Sydney, 1979.
*Buchanan, R.A. *Bush Regeneration: Recovering Australian Landscapes.* TAFE NSW, Sydney, 1989.
*Buchanan, R.A. Pied currawongs *(Strepera graculina)*: their diet and role in weed dispersal in suburban Sydney, New South Wales. *Proceedings of the Linnean Society of New South Wales* 111, 231-255, 1989.
*Buchanan, R.A. *Common Weeds of Sydney Bushland,* Inkata Press, Melbourne, 1981.
* Campbell, J.F. The valley of the Tank Stream. *Journal of the Royal Australian Historical Society* 10, 63-103, 1924.

Chapter 5: The Vegetation of Your District

The City, Inner West and Southwest

1 Sydney

1 Phillip, Arthur *The Voyage of Governor Phillip to Botany Bay* (first published London, 1789). Facsimile edn. Hutchinson Group (Australia), Melbourne, 1982.

2 Moore, C. *Botanic Gardens: Report* 29th March 1879.

3 Maiden, J.H. *Annual Report for Botanic Gardens and Domains for year 1902.*

2 Ashfield

4 Recollections of Mr C.A. Henderson — Sydney to Homebush, 1855. *Journal of the Royal Australian Historical Society* 8, 350–358, 1923.

5 *The Railway Guide of New South Wales.* Government Printer, Sydney, 1879.

3 Auburn

6 Clarke, P. & Benson, D. The natural vegetation of Homebush Bay — two hundred years of changes. *Wetlands (Australia)* 8(1), 3–15, 1988.

7 *Rookwood Necropolis: Draft Plan of Management.* Joint Committee of Necropolis Trustees, 1990.

5 Burwood

8 Lycett, Joseph *Views in Australia.* London, 1824.

6 Canterbury

9 Bradley, William *A Voyage to New South Wales: The Journal of Lieutenant William Bradley RN of HMS Sirius 1786–1792.* Facsimile edn. Trustees of the Public Library of N.S.W. & Ure Smith, Sydney, 1969.

**Canterbury Municipality Bushland Survey.* National Trust of Australia (NSW), Sydney, 1983.

7 Concord

10 Coupe, S. *Concord: A Centennial History.* Concord Municipal Council, 1983.

8 Drummoyne

11 Quoted in Allars, K. The Five Dock Farm. *Journal of the Royal Australian Historical Society* 34, 89–105, 1948.

9 Hurstville

12 *Hurstville Municipality Bushland Survey.* National Trust of Australia (NSW), Sydney, 1981.

10 Kogarah

**Kogarah Bushland Survey.* National Trust of Australia (NSW), Sydney, 1979.

11 Leichhardt

13 Cunningham, P. *Two years in New South Wales.* London, 1827.

*Dick, M. *Native Plants of Inner Western Sydney.* Leichhardt Association, Sydney, 1980.

13 Rockdale

14 Andrews, E.C. Beach formations at Botany Bay. *Journal and Proceedings of the Royal Society of NSW* 46, 158–185, 1912

15 Carruthers, Sir Joseph Captain Cook and Botany Bay: comments on the paper by Mr C.H. Bertie. *Journal of the Royal Australian Historical Society* 11, 32–38, 1925.

16 Geeves, P. & Jervis, J. *Rockdale, its Beginning and Development.* Rockdale Municipal Council, undated.

17 Robinson, L. *Trees of Wolli Creek.* Wolli Creek Preservation Society, Earlwood, 1987.

18 *Rockdale Bushland Survey.* National Trust of Australia (NSW), Sydney, 1988.

14 South Sydney

19 Mitchell, T.L. to John Mitchell, 3101828, quoted in Mourot, S. *This Was Sydney: A Pictorial History from 1788 to the Present Time.* Ure Smith, North Sydney, 1969.

20 Etheridge, R. & David, T.W.E. On the occurrence of a submerged forest, with remains of the dugong, at Sheas Creek, near Sydney. *Journal and Proceedings of the Royal Society of NSW* 30, 158–185, 1896.

15 Strathfield

21 Meredith, Mrs Charles *Notes and sketches of New South Wales* (John Murray, London, 1844). Facsimile edn. Penguin Books, Ringwood, 1973.

22 Maiden, J.H. *Agricultural Gazette of New South Wales* p 758, 1893.

Parramatta and the Cumberland Plain

16 Parramatta

23 White, John *Journal of a voyage to New South Wales* (first published 1790). Angus & Robertson, Sydney, 1962.

24 Bennett, George *Wanderings in New South Wales, Batavia, Pedir Coast, Singapore and China.* Volume I. Richard Bentley, London, 1834.

17 Blacktown

25 Hunter, John *An historical journal of events at Sydney and at Sea 1787–1792* (first published London, 1793). Angus & Robertson, Sydney, 1968.

18 Camden

26 Collins, David *An account of the English Colony in New South Wales* (first published London, 1798). Facsimile edn. A.H. & A.W. Reed, Sydney, 1975.

27 Macarthur to King 12/10/1805, *Historical Records of Australia* 5, 578.

28 Woolls, W. *A Contribution to the Flora of Australia.* F. White, Sydney, 1867.

29 Mitchell, Sir Thomas *Three Expeditions into the Interior of Eastern Australia* (first published London, 1839). Facsimile edn. Libraries Board of South Australia, Adelaide, 1965.

19 Campbelltown

**Campbelltown City Council Bushland Survey.* National Trust of Australia (NSW), Sydney, 1987.

20 Fairfield

30 Sydney [*Morning*] *Herald* 7/11/1831.

21 Hawkesbury

31 Atkinson, James *An account of the State of Agriculture and Grazing in New South Wales.* (first published London, 1836). Facsimile edn Sydney University Press, Sydney, 1975.

22 Holroyd

32 Tench, Watkin *Sydney's First Four Years: Being a reprint of A Narrative of the Expedition to Botany Bay and A Complete Account*

of the Settlement at Port Jackson. Library of Australian History, Sydney, 1979.

23 Liverpool

33 Maiden, J.H. *Forest Flora*. Volume 6. Government Printer, 1917.

*Benson, D.H. Aspects of the ecology of a rare tree species, *Eucalyptus benthamii*, at Bents Basin, Wallacia. *Cunninghamia* 1(3), 371-383, 1985.

*Benson, D.H., Thomas, J. & Burkitt, J. The natural vegetation of Bents Basin State Recreation Area. *Cunninghamia* 2(2), in press 1990.

24 Penrith

34 Phillip to Sydney 13/2/1790, *Historical Records of Australia* 1, 156.

35 Bennett, George *Gatherings of a Naturalist in Australasia* (first published London, 1860) Facsimile edn. Currawong Press, Sydney, 1982.

36 Lesson, Rene Journey across the Blue Mountains, 1824. In: *Fourteen Journeys over the Blue Mountains of New South Wales 1813-1841* (G. Mackaness, ed.). Part II. Review Publications, Dubbo, 1978.

37 Betteridge, C. Legislative protection for historic gardens. In: *Proceedings of the First Garden History Conference.* National Trust of Australia (Victoria), Melbourne, 1980.

38 Forestry Commission of N.S.W. *Management Plan for Cumberland Management Area.* The Commission, Sydney, 1984.

*Benson, D.H. Vegetation of the Agnes Banks sand deposit, Richmond, New South Wales. *Cunninghamia* 1(1), 35-58, 1981.

The Eastern Suburbs

25 Botany

39 Nicholls, M. (ed.) *Traveller under concern: The Quaker journals of Frederick Mackie on his tour of the Australian colonies 1852-1855.* Cat & Fiddle Press, Hobart, 1973.

40 Gibbs, Shallard & Co. *An Illustrated Guide to Sydney 1882.* Facsimile edn. Angus & Robertson, Sydney, 1981.

41 Beasley, M. *The Sweat of their Brows.* Water Board, Sydney, 1988.

42 Hamilton, A.A. Topographical, ecological, and taxonomic notes on the ocean shoreline vegetation of the Port Jackson district. *Journal and Proceedings of the Royal Society of New South Wales* 51, 287-355, 1918.

26 Randwick

43 Beaglehole, J.C. (ed.) *The Voyage of the* Endeavour *1768-1771.* Cambridge University Press, Cambridge, 1968.

44 Hamilton, A.A. An ecological study of the saltmarsh vegetation in the Port Jackson district. *Proceedings of the Linnean Society of New South Wales* 44, 463-513, 1919.

45 Lynch W.B. & Larcombe F.A. *Randwick 1859-1976.* Oswald Ziegler Publications, Sydney, 1976.

46 Mundy, G.C. *Our Antipodes.* Bentley, London, 1852.

27 Waverley

47 Maiden, J.H. *Some remarks on the sand-drift problem.* Miscellaneous publication No. 351, New South Wales Dept of Agriculture, Sydney, 1900.

28 Woollahra

48 Maiden, J.H. & Betche, E. Notes from the Botanic Gardens, Sydney, No. 1. *Proceedings of the Linnean Society of New South Wales* 22, 146-149, 1897.

* *Woollahra Municipality: Bushland Survey.* National Trust of Australia (NSW), Sydney, 1981.

North of the Harbour

29 Baulkham Hills

49 Burton, D. Report to Governor Phillip, 24/2/1792. *Historical Records of New South Wales* 1(2), 599.

50 *The Beginnings of the Hills District.* Hills District Historical Society, Castle Hill, 1987.

51 Proudfoot, H. *Exploring Sydney's West.* Kangaroo Press, Sydney, 1987.

52 Forestry Commission of N.S.W. *Cumberland State Forest Walking Trails.* The Commission, Sydney, (undated).

53 The *Australian* newspaper, 29/8/1827, quoted in *The Beginnings of the Hills District.*

54 Benson, D.H. Survey of the natural vegetation of the floodplain of the Hawkesbury River and its major tributaries. In: *Geomorphology of New South Wales Coastal Rivers* (H.A. Scholer, ed.). Report No. 139, Water Research Laboratory, University of New South Wales, Manly Vale, 1974.

*Clark, S. (ed.) *A Resources Study of Excelsior Park.* Report No. 86/83, Centre for Environmental and Urban Studies, Macquarie University, Sydney, 1983.

30 Hornsby

55 MacLeod Morgan, H.A. 'The Account of a Journey to the Sea in the Month of February, 1805.' Another Expedition of George Caley. *Journal of the Royal Australian Historical Society* 43, 260-266, 1958.

56 Hornsby Shire Historical Society *Pioneers of Hornsby Shire 1788-1906.* Library of Australian History, Sydney, 1979.

57 Thomas, J. & Benson, D.H. *Vegetation Survey of Ku-ring-gai Chase National Park.* Royal Botanic Gardens, Sydney, 1985.

58 Outhred, R., Lainson, R., Lamb, R. & Outhred, D. A floristic survey of Ku-ring-gai Chase National Park. *Cunninghamia* 1(3), 313-338, 1985.

59 Thomas, J. & Benson, D.H. *Vegetation Survey of Muogamarra Nature Reserve.* Royal Botanic Gardens, Sydney, 1985.

60 *A Plan of Management for Pennant Hills Park and some surrounding Bushland.* Beecroft Cheltenham Civic Trust, 1976.

61 *A guide to Elouera Bushland Natural Park.* Elouera Bushland Park Trust, Hornsby, 1983.

* Smith, P. & J. *Hornsby Shire Bushland Survey* Hornsby Shire Council, 1990.

31 Hunters Hill

62 Travis Partners *et al.*, *Kellys Bush Landscape and Management Plan: Final Report.* Report to the New South Wales Department of Environment and Planning, 1986.

63 *Hunters Hill Bushland Survey.* The National Trust of Australia (NSW), Sydney, 1990.

32 Ku-ring-gai

64 Buchanan, R.A. *Municipality of Ku-ring-gai Bushland Management Survey.* Ku-ring-gai Municipal Council, 1983.

65 A *Plan for Management for South Turramurra Bushland.* South Turramurra Environment Protection & R. Buchanan, undated.

66 *Lane Cove River State Recreation Area.* Lane Cove River State Recreation Area Trust & James Mitchell and Associates, 1983.

67 Clarke, P.J. & Benson, D.H. *Vegetation Survey of Lane Cove River State Recreation Area.* Royal Botanic Gardens, Sydney, 1987.

33 Lane Cove

68 Russell, E. *Lane Cove 1788-1970 : A North Shore History.* Lane Cove Municipal Council, 1970.

69 McLoughlin, L. *The Middle Lane Cove River : A History and a Future.* Monograph No. 1, Centre for Environmental and Urban Studies, Macquarie University, Sydney, 1985.

70 McLoughlin, L. Mangroves and grass swamps: changes in the shoreline vegetation of the middle Lane Cove River, Sydney, 1780's-1880's. *Wetlands (Australia)* 7(1), 13-24.

*Dove, D. *Bushwalks around Lane Cove.* Project Environment, Lane Cove, 1984.

34 Manly

*Clemens, J & Franklin, M.H. A description of coastal heath at North Head, Sydney Harbour National Park: impact of recreation and other disturbance since 1951. *Australian Journal of Botany* 28, 463-478, 1980.

**Manly Municipality: Bushland Survey.* National Trust of Australia (NSW), Sydney, 1981.

35 Mosman

71 Bradley, J. *Bush Regeneration.* The Mosman Parklands and Ashton Park Association, 1971.

72 Bradley, J. *Bringing Back the Bush* (J. Larking, A. Lennig & J. Walker, eds). Lansdowne Press, Sydney, 1988.

**Mosman Municipality: Bushland Survey.* National Trust of Australia (NSW), Sydney, 1978.

36 North Sydney

73 Thorne, L.G. *A History of North Shore Sydney from 1788 to Today.* Angus & Robertson, Sydney, 1979.

37 Ryde

74 Shearer, W.G. & Jenkins, B.W. (eds) *Resource Book for Field Studies in the Ryde District.* Association for Environmental Education (NSW), Sydney, 1979.

**Ryde Municipality: Bushland Survey.* National Trust of Australia (NSW), Sydney, 1982.

38 Warringah

75 Helen Veitch to Lyn McDougall, personal communication, 1988.

**McKillop Park, Dee Why Headland, Long Reef Point Bushland Survey.* National Trust of Australia (NSW), Sydney, 1985.

*The Wildlife Subcommittee *Bantry Bay: The Case for Conservation.* Upper Middle Harbour Conservation Committee, undated.

39 Willoughby

76 Fox, Allan & Assoc. *Willoughby Municipal Council Plan of Management for Bushland Reserves.* Willoughby Municipal Council, 1984.

77 Buchanan, R.A. *Mowbray Park: Description and Management.* Mowbray Park Preservation Society, Sydney, 1979.

The Southern Suburbs

40 Sutherland

78 Beaglehole, J.C. (ed.) *The* Endeavour *journal of Joseph Banks 1768-1771.* Volume 2. Angus & Robertson, Sydney, 1962.

79 Australian Littoral Society *An Investigation of Management Options for Towra Point, Botany Bay.* Australian National Parks & Wildlife Service, Canberra, 1977.

80 Cridland, Frank *The Story of Port Hacking, Cronulla and Sutherland Shire.* Angus & Robertson, Sydney, 1924.

*Maiden, J.H. *Sir Joseph Banks: The 'Father of Australia'.* Government Printer, Sydney, 1909.

*McGuiness, K. *The Ecology of Botany Bay and the Effect of Man's Activities: A Critical Synthesis.* The Institute of Marine Ecology, University of Sydney, Sydney, 1988.

*Mitchell, M.L. & Adam, P. The decline of saltmarsh in Botany Bay. *Wetlands (Australia)* 8(2), 55-60, 1989.

*Urwin, N. *A Descriptive Survey of the Native Vegetation of Kurnell Peninsula.* Working Paper 79/24, Department of Environment and Planning, Sydney, 1979.

Chapter 6: Sydney's Bushland from Now On

1 Benson, D.H. & Howell, J. Sydney's vegetation 1788-1988 — utilization, degradation and rehabilitation. *Proceedings of the Ecological Society of Australia,* 16, 115-127, 1990.

2 Benson, D.H. Maturation periods for fire-sensitive shrub species in Hawkesbury Sandstone vegetation. *Cunninghamia* 1(3), 339-349, 1985.

3 Rackham, O. *The History of the Countryside.* J.M. Dent & Sons Ltd, London, 1986.

4 Edwards, G.P. How to destroy the bush without really trying. *Landscape Australia* 2, 1989.

5 Johnson, D.L. *The Architecture of Walter Burley Griffin.* Macmillan, North Sydney, 1977.

*Buchanan, R.A. *Bush Regeneration: Recovering Australian Landscapes.* TAFE NSW, Sydney, 1989.

*Buchanan, R.A. *Common Weeds of Sydney Bushland.* Inkata Press, Melbourne, 1981.

*Corbett, J. Conservation and Management of the Cumberland Plain. *National Parks Journal* 31(3), 1987.

*Lord, S. & Daniel, G. *Bushwalks in the Sydney Region.* National Parks Association of NSW, Sydney, 1989.

*McDougall, G. & Shearer-Heriot, L. *The Great North Walk.* Kangaroo Press, Sydney, 1988.

*Maloney, B. & Walker, J. *More About Bush Gardens.* Horwitz Grahame, Sydney, 1981.

**How to Manage Your Bushland Reserves.* Nature Conservation Council of N.S.W., Sydney, 1984.

*Norrie, D. *Urban Parks, Bushland and other Natural Areas: Community Guide to Development and Planning Proposals for Public Open Space.* Total Environment Centre, Sydney, 1988.

*Norrie, D. *Urban Parks, Bushland and other Natural Areas: Guidelines for Preparing Inventories and Management Plans.* Total Environment Centre, Sydney, 1988.

*Norrie, D. *Urban Parks, Bushland and other Natural Areas: Planning, Development Control and Management — Review of Legislation.* Total Environment Centre, Sydney, 1988.

*Paton, N. *Sydney Bushwalks.* Kangaroo Press, Sydney, 1986.

Index

Local Government Area boundaries, Sydney, 1990